An Introduction to Plant Immunity

Authored by

Dhia Bouktila
*Higher Institute of Biotechnology of Béja, University of
Jendouba (Tunisia) &
Higher Institute of Biotechnology of Monastir, University of
Monastir (Tunisia)*

&

Yosra Habachi
*Independent writer
University of Tunis El Manar
Tunisia*

An Introduction to Plant Immunity

Authors: Dhia Bouktila and Yosra Habachi

ISBN (Online): 978-1-68108-802-0

ISBN (Print): 978-1-68108-803-7

ISBN (Paperback): 978-1-68108-804-4

Published by Bentham Science Publishers – Sharjah, UAE. All Rights Reserved.

need for a court order if at any point you breach any terms of this License Agreement. In no event will any delay or failure by Bentham Science Publishers in enforcing your compliance with this License Agreement constitute a waiver of any of its rights.

3. You acknowledge that you have read this License Agreement, and agree to be bound by its terms and conditions. To the extent that any other terms and conditions presented on any website of Bentham Science Publishers conflict with, or are inconsistent with, the terms and conditions set out in this License Agreement, you acknowledge that the terms and conditions set out in this License Agreement shall prevail.

Bentham Science Publishers Ltd.
Executive Suite Y - 2
PO Box 7917, Saif Zone
Sharjah, U.A.E.
Email: subscriptions@benthamscience.net

CONTENTS

FOREWORD

It has often been said that plants have neither nervous nor immune systems. However, they are able to react to certain stimuli in the environment and to different stresses whether they are biotic or abiotic. They can fight or resist the germs that surround them. In fact, if they do not have an immune system *per se*, neither dedicated organs nor immune cells, plants have defense mechanisms that can be compared to those of so-called innate immunity in the animal kingdom.

First, they have natural physical (cuticles or spines) or chemical (wax or other compounds) barriers that allow them to prevent or limit infections and control pests. From this point of view, they use the same passive defense strategies as animals (skin, mucous membranes, sweat, sebum, acid secretions).

When pathogens cross these barriers, they encounter active defense mechanisms at the cellular level, with the same molecular systems for the perception of microbial aggressions. These systems involve surface and intracellular receptors PRRs (Pathogen Recognition Receptors) that recognize PAMPs (pathogen-associated molecular patterns) or DAMPs (damage-associated molecular pattern molecules) and trigger signaling pathways aimed at carrying out resistance to infection. Interestingly, these recognition molecules are also described for innate immunity cells in animals. While PRRs identified in plants are primarily, in cell membrane, mammalian receptors can be membrane, cytoplasmic or localized in the endosome membrane. This recognition displays certain specificity and there is a diversity of PRRs recognizing PAMPs, which are conserved patterns, common to different germs and pathogenic microorganisms. Some bacterial PAMPs, such as flagellin (Flg), lipopolysaccharides (LPS) and peptidoglycans (PGN) are recognized in both plants and animals. However, if PRR orthologs in animals do not appear to exist in plants, protein domains such as LRRs (leucine-rich repeats) are conserved between the PRRs of the two kingdoms. For example NOD2 (nucleotide-binding oligomerization domain 2), a cytoplasmic PRR whose mutations are associated with Human Crohn disease, is homologous to resistance *R* proteins in plants.

Similarities between the two kingdoms are also observed in reaction processes including signaling and immune responses. In animals and plants, the perception of PAMPs and DAMPs induces a signaling cascade that leads to the activation of transcription factors and results in the transcription of defense genes. Signal transduction in plants and mammals also involves altering ion flow through membranes, especially Ca^{2+}, as well as producing ROS (reactive oxygen species) or NO (nitrogen monoxide). All of these second messengers also contribute to the expression of a defense transcriptome.

In both cases, pathogen-activated cells have a reprogrammed transcription profile associated with transcription factors (TFs) induced *via* the signaling pathways. These TFs regulate key genes of the protective response. In plants, the target genes, encode in particular, enzymes involved in the synthesis of phytohormones that amplify the immune response and warn neighbouring cells. This is to be compared with what is observed in animal models in which the recognition of PAMPs by TLRs, for example, leads to transduction pathways regulating the expression of genes encoding mediators of inflammation (*e.g.* cytokines and chemokines) that allow the recruitment of specialized defense cells. Moreover, plants and animals synthesize antimicrobial compounds, some of which are common to both kingdoms, such as those of the defensin and thionin family.

In some cases, plant immunity can lead to programmed cell death, called hypersensitive response, which helps to limit the spread of microorganisms. It results in the appearance of localized necrotic lesions at the sites of infection. The mechanisms leading to this cell death can be compared in some ways to the apoptosis or pyroptosis that is observed in animals. Several events of programmed cell death are indeed similar between the two kingdoms.

The adaptation of pathogens to their host is essentially manifested by bypassing the host's immunity. Microbes that infect animals use a variety of escape strategies to reduce the host's defenses and infect them. As in animals, pathogens that infect plants are able to manipulate the cellular functions of their host through effectors (proteins, toxins, *etc.*), thus facilitating their spread. These effectors largely target the cellular signaling pathways that lead to the immune response but also those that induce the opening of the stomata.

Another aspect known for several years in plants is cross-protection. This aspect has been described, in the last decade, in vertebrates and explained by a mechanism known as innate memory or trained immunity. Indeed, it is now accepted that activated innate immunity cells can be maintained in their state of activation for several months with a reprogramming of the transcriptional profile determined, not only by transcription factors, but also by epigenetic changes in DNA and histones methylation profiles, as well as in microRNA expression. Thus, the cells of innate immunity induced into memory cells will persist under an activation state for several months during which they will be more receptive and respond to other infections more effectively. Hence is explained the non-specific cross-protection that can be observed in animals and also in plants where epigenetic change associated to immune response are documented.

Thus, plants use defense strategies against biotic aggressions, which are very similar to those of vertebrates. Knowledge of the mechanisms of innate animal immunity could, therefore, guide research to a better understanding of the defense pathways induced by PAMPs and DAMPs in plants. At a time when agro-ecological concerns are guiding us towards reducing inputs in agriculture, this knowledge will lead to the development of new biocontrol strategies based on stimulating the natural defenses of plants.

This book is a basic document for those who want to understand and deepen their knowledge of immunity in plants. All aspects are dealt with in a gradual and clear way, including the basic concepts, and the subject is treated from its multiple facets: microbiological, phytopathological, cellular, biochemical, genetic, evolutionary, biotechnological and agronomic.

Amel Benammar Elgaaïed
Tunisian Academy of Sciences
Tunisia

PREFACE

The world population is increasing day by day, imposing that agricultural production increases proportionally. Unfortunately, the increase in agricultural production is limited due to abiotic and biotic stresses, which have a negative impact on the growth and development of crops and their yields, resulting in considerable economic loss worldwide. In addition, monoculture, which we are witnessing today in agricultural landscapes, and which represents the workhorse of intensive agriculture, weakens plants enormously, while biodiversity protects them. It should be noted that the economic loss attributable to plant diseases and pests, which is estimated at 10-40% of the production potential worldwide, is already being held back by the massive use of phytosanitary products. Although this use of chemicals helps plants to fight against pathogenic microorganisms, it is not without consequences for the environment.

At the sunrise of this new millennium, mankind has an extraordinary chance to enhance development through the scientific improvement of crops and sustainable management of biodiversity. Indeed, advances in Genetics, Molecular Biology and Genomics have evolved to allow what is now called a ***high-tech plant breeding***. Improvements in plant performance can be manifested, among other things, by ***tolerance or resistance to abiotic and biotic stresses***. It follows all the importance that the acquisition of a deep understanding of the genetic mechanisms governing the response of a plant to its invasion by pathogens and pests currently takes. Plants have their own mechanisms for overcoming stress, and activating complex - and sometimes crossed - signal transduction pathways. In recent years, knowledge of the immunogenetic defense mechanisms of plants has improved considerably.

In this perspective, we wanted this book to be an extensive and up-to-date scientific review of the different aspects of knowledge and technologies related to the field of plant immunity. Our goal is to offer a real modern tool for learning and documenting plant immunity for specialist academicians. We hope that students from all over the world will be able to benefit from this informative resource, while allowing teacher-researchers to use it for their teaching and research activities.

Dhia Bouktila
Higher Institute of Biotechnology of Béja
University of Jendouba (Tunisia) &
Higher Institute of Biotechnology of Monastir
University of Monastir (Tunisia)

Yosra Habachi
Independent writer
University of Tunis El Manar
Tunisia

ACKNOWLEDGEMENTS

This work was facilitated by a one academic year teaching leave awarded to Dhia Bouktila, Associate Professor at the Higher Institute of Biotechnology of Beja, in order to focus on completing the writing of this book and achieving this publication. For this, the authors thank the Scientific Council of the Higher Institute of Biotechnology of Beja, the University of Jendouba Council, as well as the Tunisian Ministry of Higher Education and Scientific Research.

The authors want to thank everyone who helped them create this book, especially the students Jihen Hamdi for her valuable help in the preparation of chapters 8 and 14, and Narjess Kmeli and Inchirah Bettaieb for their help in formatting bibliographical entries. Special thanks goes to our editor, **Bentham Science Publishers** who welcomed us into this publishing experience and assisted us with great professionalism.

The authors disclose receipt of the following financial support for the authorship and publication of this book. This work was supported by the Tunisian Association of Psycho-Neuro-Endocrino-Immunology (Association Tunisienne de Psycho-Neuro-Endocrino-Imm unologie) [grant number 2-2020].

CONSENT FOR PUBLICATION

Not applicable.

CONFLICT OF INTEREST

The authors declare no conflict of interest, financial or otherwise.

DEDICATION

To my wife, Rafika, and my kids, May and Marwen: Your support made all of this possible,
To my parents for always loving and supporting me,
To my dear friend and coauthor, Yosra Habachi,
To every student I have taught or advised once in my carreer,
A special dedication to dear friends, Dr. Chokri Belaid, Dr. Anwar Mechri, Dr. Khaled Chatti, Dr. Lotfi Cherni and Dr. Salim Lebbal.

Dhia Bouktila

To my husband, Sami, for his irreplaceable support and his unconditional love.
To my little angels: Seif Allah and Chahed, as a testimony to the attachment, love and affection I have for them.
To my dear friend Dhia Bouktila,
To Dr. Abdelfatteh Zeddini, Head of Laboratory of Pathological Anatomy and Cytology, for his kindness and support.
To my dear colleagues, Asma and Yathreb, for their loveliness and the help that each of them has given me.

Yosra Habachi

CHAPTER 1

Introduction

The survival of most organisms under various environmental conditions depends on the presence of general immune mechanisms, governed by an integrated genetic system. Plants, despite their immobility, have developed various sophisticated and effective mechanisms to recognize and combat pathogens during their attacks. Plant immunity is defined as ***the ability of plants to contain the damaging effect of a pathogen or pest***. Plants contain the genetic information necessary to defend themselves from attack by a multitude of plant pathogens and pests such as viruses, bacteria, insects, nematodes, fungi and oomycetes. This defense can operate at different levels, using either preexisting passive defense systems (cuticle, wax, thorns, chemical compounds, *etc.*), or active defense systems appearing after the perception of aggression. In most cases, the first line of defense is sufficient to repel the pathogen, but sometimes the constitutive barriers are not sufficient and the second, active, line of resistance will be required.

Cell wall penetration introduces microbes to the plant plasma membrane where they will be confronted with extracellular surface receptors that detect pathogen-associated molecular patterns (PAMPs). This detection of microbes on the cells surface sets up ***PAMP-triggered immunity (PTI)***, which hopefully prevents the infection well before the pathogen begins to spread in the plant. That being said, pathogens have evolved strategies to disrupt PTI by secreting specific proteins, called effectors, in the cytosol of plant cells, which affect the efficacy of primary resistance (PTI). Once pathogens have gained the potential to eliminate primary defenses, plants, on the other hand, will establish a more advanced framework for the detection of microbes, termed ***effector-triggered immunity (ETI)***. In the scenario of ETI, the products of major ***resistance (R) genes***, normally intracellular receptors, perceive the associated effector molecules released by the pathogen inside the host cell. Interplay between effectors and intracellular receptors activates a dynamic signaling network to gain disease resistance (McDowell and Dangl 2000). In fact, plant disease resistance conferred by *R* genes is usually supported by an ***oxidative burst***, which is a rapid generation of significant amounts of reactive oxygen species (ROS). This ROS output is necessary for another component of the resistance process, called *hypersensitive response* (HR), a form of programmed cell death that is assumed to restrict pathogen access to the plant.

Finally, at the molecular level, the plant coordinates *the transcription of a variety of genes* whose sole objective is resistance. The success of the plant depends on the intensity and speed of the perception of the pathogen signals and their transmission in it to produce an effective response against the pathogen. In *Arabidopsis*, the identification of pathogen-responsive genes is the subject of numerous studies. It has been found that no less than 25% of the genes identified in this model plant species have a transcriptional level affected following the attack of a pathogenic agent. In this way, a deeper knowledge of the basic processes involved in defense responses would make it easier to interpret the interactions between plants and pathogens and allow better resistance of plants, especially in species of agronomic interest.

The relationship between a plant and a harmful organism (*i.e.* pathogen or pest) depends on the environmental conditions, the properties of the harmful organism and the plant's ability to defend itself. The concomitant evolution at the genetic level, including the plant and its pathogenic organism, is a *coevolutionary process*, which means a specific reciprocal interaction, between the plant and the pathogen. It obviously follows that a large part of the diversity of the living world comes from this coevolution between plants and pathogens, which seems to be an interminable arms race: a species induces a behavioral response to selection pressure imposed by another antagonistic species and the latter changes its behavior in response to the change in the first species. In all coevolutionary systems, the two partner species seek to stabilize with a balanced genetic structure. However, the structure of the genomes of any living organism is constantly modified according to the evolutionary race, *via*, both, small (point mutations) or large-scale (whole-genome duplications) events.

When a pathogen colonizes a plant, or a pest chooses it as a food resource, this will exert a *selection pressure* on the plant, thus reducing its fitness. The plant will react in two ways, either it definitively eliminates the aggressor; it is, in this case, a resistant plant, or it accepts the invasion by activating a compensation process; it is, in this case, a tolerant plant. Thus, in-depth knowledge of the genetic defense mechanisms involving resistance genes against biotic stress in plants is a prerequisite for the implementation of management programs and effective control, taking into account of the concomitant evolution of the two protagonists involved.

Contrary to popular belief, the first study linking the development of a disease to a microorganism was not carried out by Robert Koch on the tuberculosis Bacillus in 1890. Instead, at the beginning of the 19th century, the cause of wheat decay was identified by the Swiss Isaac-Bénédict Prévost (1755-1819). This researcher analyzed the cycle of the microscopic parasite responsible for this disease, and developed a mixture capable of eradicating it. However, this work was forgotten because of the preference, in official scientific circles, for the theory of ***spontaneous generation[1]***. In 1861, the German Anton de Bary, considered as the father of phytopathology, did the same by proving that the terrible epidemic of potato late blight responsible for the great famine of Ireland of the 19th century was caused by the filamentous pathogen *Phytophtora infestans* (Matta 2010). More recently, the fungus *Helminthosporium oryzae* was the cause of one of the most significant famines of the 20th century. In 1943, the destruction of rice crops by this fungus was responsible for the deaths of three million people in Bengal (Padmanabhan 1973). The practice of intensive farming since the late 1970s encouraged the development of epidemics. Indeed, monoculture on very large plots and the shortening of crop rotations have led to a loss of diversity in cultivated plants, which are no longer able to resist pathogenic agents on a long-term basis (Ricci *et al.* 2011). Control of phytopathogenic agents is, therefore, a major issue to ensure food security for populations.

To limit the damage caused by pathogens in agrosystems, humans have developed various control methods. First of all, cultural practices make it possible to limit the quantities of inoculum, by crop rotation and the burial of residues. Chemical control has also been widely used since the start of the 20th century and has significantly increased yields (Hirooka and Ishii 2013). Chemical treatments of crops effectively fight against phytophagous insects and fungal diseases. However, their possible impact on the environment is a real source of concern. In addition, as it is the case in animals and humans, chemical treatments are powerless against viral plant diseases, except in the rare cases where they attack the organisms that vector them, insects, nematodes or fungi. It is therefore necessary to develop alternative strategies to chemical control, against viruses. Finally, genetic selection is based on the use of ***cultivar resistance*** to fight against pathogens.

Two types of cultivar resistance are differentiated. ***Quantitative resistance*** is controlled by a large number of genes (polygenic resistance) associated with genome portions called Quantitative Trait Loci (QTLs) that contribute to the expression of resistance. It most often gives the plant partial resistance to a pathogen because the defenses of the plant do not completely prevent the invasion of the disease. It is therefore not blocked but only slowed down in its progression, which causes some visible damage to the plants. On the other hand, ***qualitative***

resistance is controlled by one or a few genes called *R* genes (mono- or oligogenic control). It most often gives the plant complete resistance to a pathogenic agent. The latter is blocked from the early stages of infection and does not cause damage. This resistance is often associated with symptoms of hypersensitivity (HR) and triggered by molecules manipulating the structure or cellular functions of the host, called effectors or virulence factors (Greenberg and Yao 2004).

In this context, the genetic dissection of quantitative resistance to diseases, *a priori* more durable than monogenic resistance, has progressed considerably over the past fifteen years with the development of ***molecular tools*** and ***genomics***. However, even if numerous studies on quantitative resistance in various pathosystems have made it possible to identify QTLs of resistance, their exploitation in cultivar breeding remains difficult because of the complexity of genetic determinisms and the instability of their effects, partly due to their interactions with the genetic background.

[1] Theory stating that life may arise from nonliving matter.

CHAPTER 2

Plant Pathogens and Plant Pests

Abstract: The diversity of plant aggressors is impressive since it includes cellular or sub-cellular pathogens (fungi, bacteria, viruses, mycoplasmas), weeds, animals (rodents, snails), and insects. Before starting a fight against plant disease, it is necessary to identify the pathogen responsible and to know its ecology, its life cycle and its mode of dissemination in the environment. The constant identification of new taxa is continuously accompanied by a revision of the classifications thanks to the new tools brought by molecular biology. In this chapter, the most recent knowledge on classes of plant pathogens and plant pests in relation with their plant hosts is presented.

Keywords: Arthropods, Bacteria, Diversity, Fungi, Host range, Mycoplasmas, Nematodes, Oomycetes, Pathogens, Parasitic Plants, Pests, Viruses, Viroids.

INTRODUCTION

Plants are the hosts of many living things that can be harmful to them. A pathogen is a biological agent responsible for an infectious disease. This definition includes fungi, oomycetes, bacteria, viruses, viroids, mycoplasmas, protozoa, nematodes and parasitic plants. From this definition are excluded ectoparasites which affect plant health by eating plant tissue. These organisms are considered as ***pests***. Pests include arthropods (such as insects and mites), as well as slugs and snails.

In the context of pathogens, the term ***parasites*** is generally reserved for microorganisms - fungi, oomycetes, bacteria, mollicutes (bacteria-like organisms lacking a rigid cell wall) - and for viruses and viroids, even if ***phytophagous agents*** (nematodes, insects and other pests) and parasitic plants can also cause significant damage to crops. An approximate assessment reports at least 8,000 species of filamentous microorganisms (fungi and oomycetes), about 200 species of bacteria and mollicutes, and more than 500 viruses and viroids.

On the other hand, ***crop pests***, also called pests, are animal organisms that attack cultivated plants, or stored crops, causing economic damage to the detriment of farmers. Pests can cause direct damage to cultivated plants by their diet (phytophagous, xylophagous, *etc.*) or their parasitic lifestyle, or indirect when they are vectors of diseases (*e.g.* viruses).

Phytopathogenic organisms or pests belong to several taxonomic classes, of which we can cite:

1. SUBCELLULAR PATHOGENS

1.1. Viruses

Diseases caused by viruses pose a serious threat to different cultures every year in many parts of the world. Currently, around 1,000 viruses infect different plants. The visible symptoms caused by viruses in plants can vary depending on the virus, the variety or species affected, and the physiological state of the plant. Many viruses cause symptoms of mosaic on the foliage which can be associated with deformations (thread-like or embossed appearance, reduction in size, *etc.*). Other viruses cause yellowing of the leaves. Finally, certain viruses induce more or less generalized necrosis on the leaves, flowers, fruits or stems.

Generally speaking, viral diseases reduce the growth and therefore the overall production potential of a plant.

1.2. Viroids

Viroids are the smallest infectious pathogens known, consisting merely of a short circular RNA without protein coats. Currently, there are 30 recognized viroid species ranging from 170 to 450 kb, which all infect plants, with some causing diseases while others are harmless (Foster and Fermin 2018). They replicate autonomously when introduced into host cells and their replication depends on the host's cellular machinery. Their mode of action is still unknown and it is assumed that they cause interference in the metabolism of cellular RNAs. Viroids generate dwarfism and deformation. Their transmission is mainly mechanical. Viroids are classed in only two families: ***Pospiviroidae*** and ***Avsunviroidae*** (Wilson 2014) Among the viroids identified in plants, we can indicate *Cadang-Cadang coconut viroid* (*CCCVd*), *Potato spindle tuberviroid* (*PSTVd*), *Tomato Chloric Dwarf* and *Apple Fruit Crinkle*.

2. CELLULAR PATHOGENS

2.1. Mycoplasmas (also called Mollicutes)

Mycoplasmas are placed in a separate class, ***Mollicutes***, which removes them from bacteria. Their cellular organization does not differ from that of other prokaryotes (Cacciola *et al.* 2017). However, it is recognized that this group of organisms is phylogenetically related to Gram-positive bacteria (Garnier *et al.* 2001). Their main characteristics are:

- They lack a cell wall.

- Their plasma membrane lacks peptidoglycan but contains sterols (*e.g.* cholesterol). They are resistant to penicillin and other antibiotics that act on peptidoglycans.

- A small size of their genomes.

- The low guanine (G) plus cytosine (C) content of their genomic DNA.

Plant pathogenic mycoplasmas are responsible for several hundred diseases and belong to two groups:

a. Phytoplasmas

The phytoplasmas (previously called ***MLOs: mycoplasma-like organisms***) represent the largest group of mycoplasmas, which was discovered first. In fact, until 1967, some diseases were attributed to viruses, although no viral particles could be detected in diseased plants. In 1967, thanks to the advent of electron microscopy, specialists (Doi *et al.* 1967) identified, in the phloem of diseased plants [1], polymorphic structures that were absent in the phloem of healthy plants. With the difference that they have no cell walls, the structures observed showed similarity to prokaryotic microorganisms, the mycoplasmas, which have long been known in animals as infectious agents or as saprophytes on mucous membranes. Therefore, they were called mycoplasma-like organisms (MLOs). Following this discovery, many diseases of inexplicable origin have been attributed to mycoplasma-like organisms (MLOs).

b. Spiroplasmas

Only three plant pathogenic spiroplasmas are known today; (a) *Spiroplasma citri*, the agent of citrus stubborn, was discovered and cultured in 1970 and shown to be helical and motile, (b) *S. kunkelii* is the causal agent of corn stunt, and (c) *S. phoeniceum*, responsible for periwinkle yellows, was discovered in Syria (Garnier *et al.* 2001; Cacciola *et al.* 2017).

Due to their organization, plant pathogenic mycoplasmas can only evolve in certain specific cellular niches of the host, mainly in the phloem cells. Phytopathogenic mycoplasmas are transported from a sick plant to a healthy plant by phloem-feeding insects (aphids, leafhoppers, psyllids, *etc.*). Particularly, insects of the Homoptera family, such as leafhoppers, are frequently involved. The leafhoppers, by sucking the elaborated sap from a diseased plant, also collect the mycoplasmas found there. Transmission of a mycoplasma by leafhopper is of the circulative[2] and persistent[3] type.

2.2. Bacteria

In plant tissues, phytopathogenic bacteria multiply locally in the intercellular spaces of the mesophile (apoplast), or in vascular bundles (xylem or phloem). They are confronted with the whole defense mechanism of the plant, often effective in stopping the disease. Bacteria have therefore developed effective strategies to manipulate the processes of the plant cell to their advantage and induce disease.

According to their genomic characteristics and their modes of nutrition, there are two groups of phytopathogenic bacteria: ***brute force pathogens***, responsible for soft rot (oily stains: destruction of underlying tissues), and ***stealth pathogens***, which cause leaf spots, necrosis or wilting. In the first case, the strategy of bacteria consists in degrading the cell walls by secreting enzymes into the medium *via* a type-2 secretion system (dynamic assembly of 12-15 proteins, which allows the secretion of multipartite holotoxins and hydrolases across the membrane). In the second case, the strategy consists in modulating the biological functions of the host cell, in particular its immune defenses, using a multitude of effector proteins directly injected into the cell *via* a type-3 secretion system (one polymerized protein, by which effectors are translocated directly into the cytoplasm of the target cell). However, certain phytopathogenic bacteria cannot be classified in these two groups. These bacteria do not cause rots, have no cell-wall degrading enzymes, and lack the type-3 secretion system (Holeva *et al.* 2004). They are, moreover, exclusively limited to the xylem vessels of their host plant, unlike most phytopathogenic bacteria, capable of colonizing several ecological niches. This specific adaptation is often justified by the fact that these bacteria have a reduced genome.

2.3. Fungi

Fungi are the main cause of plant disease and are responsible for around 70% of crop plant diseases. The number of fungal or pseudo-fungal organisms capable of infecting plants is estimated between 10,000 and 15,000 species. Unlike mollicutes and bacteria, fungi are capable of directly infecting healthy intact plants (Birch *et al.* 2008).

Depending on the trophic relationship between the plant and the phytopathogenic fungus, the latters can be classified into three major groups:

- **Biotrophs** (*Compulsory Parasites*): develop in close association with living cells. Specialized hyphae (haustorium) penetrate the cell wall and induce invagination of the cell membrane. Examples of biotrophic pathogens include the rust fungi (*Basidiomycota*) and the powdery mildew fungi (*Ascomycota, e.g. Blumeria graminis*).

- **Necrotrophs**: produce enzymes or toxins to kill the cells of the host plant in front of the area of infection. The hyphae develop between dead cells. This is the case, for example, of the gray mold fungus *Botrytis cinerea*.

- **Hemibiotrophs**: exhibit characteristics of both biotrophs and necrotrophs. They initially develop as a biotroph in association with living cells, but become more or less quickly necrotrophic. For example, agents of cruciferous neck necrosis (for example, fungal pathogens of the phylum *Ascomycota* that are the causal agents of blackleg disease on *Brassica* crops; *e.g. Leptosphaeria maculans* anamorph *Phoma lingam*).

The following Table **1** gives some examples of necrotrophic and biotrophic fungal pathogens, and includes some hemibiotrophic pathogens.

Table 1.   Examples of necrotrophic, biotrophic and hemibiotrophic fungal pathogens.

Necrotrophs		Biotrophs	
Pathogen	**Disease**	**Pathogen**	**Disease**
Botrytis cinerea	Grey mould	*Blumeria (Erysiphe) graminis*	Powdery mildew
Cochliobolus heterostrophus	Corn leaf blight	*Uromyces fabae*	Rust
Pythium ultimum	Damping off in seedlings	*Ustilago maydis*	Maize smut
Ophiostoma novo-ulmi	Dutch elm disease	*Cladosporium fulvum*	Tomato leaf mould
Fusarium oxysporum	Vascular wilt	*Puccinia graminis*	Black stem rust of cereals
Sclerotinia sclerotiorum	Soft rot	*Phytophthora infestans*	Potato late blight
Hemibiotrophs			
Cladosporium fulvum, causing tomato leaf mould (also called a biotroph). *Colletotrichum lindemuthianum*, causing anthracnose. *Magnaporthe grisea*, causing rice blast (also called a necrotroph). *Phytophthora infestans*, causing potato late blight (also called a biotroph by some, necrotroph by others). *Mycosphaerella graminicola*, causing *Septoria* leaf blight.			

2.4. Oomycetes

Oomycetes are eukaryotic organisms that superficially resemble filamentous fungi, but are phylogenetically related to diatoms and brown algae in the stramenopiles (Thines and Kamoun 2010). Kamoun *et al.* (2014) carried out a survey of 62 scientists in 15 countries, which yielded a ranking of top 10 species of plant-pathogenic oomycete taxa based on scientific[4] and economic importance (Table **2**).

Table 2. Top 10 oomycetes in molecular plant pathology (simplified from Kamoun *et al.* 2014).

Rank	Species	Common disease name(s)
1	*Phytophthora infestans*	Late blight
= 2	*Hyaloperonospora arabidopsidis (syn. Peronospora parasitica)*	Downy mildew
= 2	*Phytophthora ramorum*	Sudden oak death; Ramorum disease
4	*Phytophthora sojae*	Stem and root rot
5	*Phytophthora capsici*	Blight; stem and fruit rot; various others
6	*Plasmopara viticola*	Downy mildew
7	*Phytophthora cinnamomi*	Root rot; dieback
= 8	*Phytophthora parasitica*	Root and stem rot; various others
= 8	*Pythium ultimum*	Damping off; root rot
10	*Albugo candida*	White rust

The '=' sign before the ranking indicates that the species tied for that position.

2.5. Nematodes

Nematodes are unsegmented cylindrical vermiform organisms occupying very diverse ecological niches on the planet. Nematodes have very diverse diets. The species can be bacteriophagous (*e.g. Caenorhabditis elegans*), entomopathogenic (*e.g. Steinernema* spp. or *Heterorhabditis* spp.), animal parasites (*e.g. Ascaris* spp., *Brugia* spp. or *Trichinella* spp.), or even predators (*e.g. Mononchus* spp.). Among all the nematode species described, only 15% are plant parasites (Blanchard 2007). Plant-parasitic nematodes are major plant pathogens in agriculture.

2.6. Parasitic Plants

Generally, plants are characterized by their autotrophy, which is the ability to synthesize organic molecules from simple elements (CO_2, H_2O, light). These syntheses are made through photosynthesis which takes place in the chlorophyllian organs. However, during their evolution, some Phanerogam plants (about 4,000 species) have lost their autotrophy and have become parasites of other higher plants now called host plants, from which they get the nutrients they need. Parasitism has caused, in parasitic plants, the development of a particular organ, the sucker or haustorium, which represents a structural and physiological bridge allowing the transit of nutrients from the host to the parasite. The sucker also allows the parasite to be attached to the host.

Parasitic plants represent a real danger for many food crops and woody species in temperate and tropical areas. Their parasitic lifestyle causes often considerable yield losses. The following examples can be given:

- European mistletoe (*Viscum album* L.): an epiphytic parasitic plant. It grows on many trees, among which poplars and apple trees are the most susceptible. It can also attack conifers.

- Striga, commonly known as witchweed: a root parasitic plant very common in Africa, attacking food crops (sorghum, millet, corn).

- Orobanches: epirhizal (root-parasitic) holoparasitic plants, which are a real scourge for industrial crops in Europe [tobacco, rapeseed, hemp (*Cannabis sativa*), sunflower] and legumes (beans, chickpeas, lentils) in North Africa and the Middle East.

3. ARTHROPODS

Arthropod pests include insects and mites.

3.1. Insects

Insect herbivores (pests) are very diverse. They cause either direct damage (the consequence of feeding insects, both adults and larvae), or indirect damage (virus transmission, excretion of honeydew that causes sooty mold formation, and very often the two together).

3.2. Mites

Mites differ from insects, as the adults have four pairs of legs (*versus* six for insects) and lack antennae. The mites that feed on plants have rasping and sucking mouth parts that damage plants and they also transmit plant pathogens as they feed. Both thrips and mites are very small and, as a result, often avoid detection until the plant growth is visibly affected.

4. THE CONCEPT OF HOST RANGE

The host range of a parasite has a central influence on disease emergence and proliferation (Woolhouse and Gowtage-Sequeria 2005). There is indeed a clear distinction between *specialist parasites* that can colonize only one or a few genetically similar host species, and generalists that can infect a number that could exceed a hundred unrelated host species (Woolhouse *et al.* 2001; Barrett *et al.* 2009). Since many scientifically and economically important issues require inventorying and/or predicting the likely host range of plant pathogens, there are currently several databases of pathogen–host range [5].

The term *Host selectivity* can be defined as the pathogen selectivity among local plant species. Knowledge of host selectivity is useful for the study of plant disease epidemics, the estimation of the parasite biodiversity, the management of agricultural and forest systems, as well as for analyzing the risks linked to the movement or the transportation of plants and pathogens (Gilbert and Webb 2006). *Host specialization*, when pathogen or pest lineages evolve to infect a smaller range of hosts, often occurs in pathosystems and can be motivated by a *landscape change* resulting in the parasite sharing habitat with only a restricted number of its potential hosts (Navaud *et al.* 2018). However, there are also examples of *parasite adaptations* that limit the use of co-occurring potential hosts. For example, Liao *et al.* (2016) reported that some isolates of the rice blast fungus *Magnaporthe oryzae* are capable of infecting *Oryza sativa* cv. japonica varieties but not cv. indica varieties co-occurring in the Yuanyang region of China. It was suggested that specialization results from the loss of traits that are not required to infect a particular host, such as the loss of lipid synthesis in parasitoid wasps (Visser *et al.* 2010). Thus, from an evolutionary point of view, specialization is seen as a dead end (Moran 1988), since gene losses are often irreversible, preventing the transition back to *generalism* (Day *et al.* 2016). However, there is evidence of transitions from specialist to generalist parasitism (Johnson *et al.* 2009; Hu *et al.* 2014). Host range is a phenomenon governed not only by the evolutionary history of a pathogen, but also by that of its potential hosts (Poulin and Keeney 2008).

CONCLUSION

Plant pathogens and plant pests affect food crops, causing severe loss in terms of economics and production in agriculture sector, and threatening food security. So, the crucial step toward disease management under natural field conditions is to appropriately detect these bioaggressors. Plants respond to their enemies with a powerful immune system including an arsenal of chemical and signaling molecules to rid themselves of the microbes.

[1]Doi *et al.* (1967) examined **petunia** plants with aster yellows, **Paulownia** trees and **potato** plants with witch's broom disease, and **mulberry** trees with dwarf disease. In all diseased plants they found mycoplasma-like bodies in the phloem. Such bodies were absent from healthy plants.

[2] The pathogen (*e.g.* viruses, mycoplasmas) circulates through the insect vector gut, hemolymph and salivary tissue membranes to reach the salivary glands for transmission. Circulative mode can be replicative (with replication and systemic invasion of vector insect tissues) or non-replicative (without replication in vector tissues).

[3] The infectious pathogen is harboured inside the vector all the time and can even be inherited by the vector progeny.

[4] The number of papers published in 2005–2014 is based on searches of the Scopus database (http://www.scopus.com) using the species names as a query.

[5] We can cite as examples the following: The Pathogen-Host Interactions database (PHI-base)(http://www.phi-base.org/) and the Australian Plant Pest Database (https://appd.ala.org.au/).

CHAPTER 3

Plant Diseases

Abstract: Broadly defined, disease is any physiological abnormality or significant disruption in the normal health of a plant that changes its appearance or function. Due to viruses, bacteria or fungi and favored by certain environmental conditions (nutrient deficiency, soil degradation, water problems, climate change, *etc.*) and certain pests (sometimes playing a role of vector), plant diseases are sources of considerable economic losses for agriculture and forestry, and as such studied by plant pathology or phytopathology.

Keywords: Controlling methods, Diagnosis, Economic impact, Non parasitic plant disease, Parasitic plant disease, Symptoms.

INTRODUCTION

Studies of the very different accidents or diseases affecting plants during their growth or post-harvest, in addition to the analysis of the alterations of their products, constitute Plant Pathology. If ***biological agents*** are, actually, responsible for many abnormalities, further study of ***plant physiology*** shows that a large number of pathological symptoms are initially caused by a disturbance of different vegetative or reproductive functions, independently of any parasitic cause. Under these conditions, the disease diagnosis becomes very delicate, and the understanding of pathological phenomena will only be total if it takes into account a range of factors relating either to the plant itself (anatomy, physiology, biology), or to the environment where it lives (climate, soil, parasites).

Typically, research leading to a successful (*i.e.* specific and sensitive) diagnosis of a plant pathogen, will be followed by researches aiming to characterize genetically (or genotype) different pathogen races/strains. These studies are crucial for understanding the mechanistic basis of ***why a certain pathogen causes disease in one host plant and not in another*** and also ***why a certain pathogen race causes disease in one host plant cultivar and not in another***. Answering these intriguing questions is a valuable key toward the improvement of methods for controlling plant diseases, especially breeding for cultivar resistance.

1. DEFINITION OF A PLANT DISEASE

A plant disease can be defined as a succession of invisible and visible responses of the cells and tissues of a plant, following the attack of a microorganism or the modification of an environmental factor, which causes disruption of the form, function or integrity of the plant. These responses can induce a partial alteration or even the death of the plant or of some of its parts.

In the case of parasitic diseases, it is noteworthy that the term disease will exclude plant injuries caused by insects: A plant is diseased when it is *continuously* affected by a biotic factor, which causes a disorder in its normal structure or activities, showing some outward signs and/or symptoms of disease. The word *continuously* excludes such things as insect injury. Disease is a term usually reserved for those problems caused by parasites: fungi, bacteria, viruses, mycoplasma, **nematodes** and parasitic higher (seed) plants.

2. CLASSIFICATION OF PLANT DISEASES

Plant diseases are caused by agents that are both infectious (fungi, bacteria, viruses and nematodes) and non-infectious (mineral deficiency, sunburn, *etc.*). Infectious diseases of plants are caused by living organisms that attack and feed on the plant they infect. The parasitic organism that causes the disease is called a *pathogen*, and the plant invaded by the pathogen and used as a nutrient source is designated as a *host*. A supportive environment is of crucial importance for the development of the disease. Even if susceptible plants are exposed to huge amounts of a pathogen inoculum, they may not develop the disease unless the environmental conditions are favorable.

Thousands of diseases affect plants. On average, each plant can be affected by a hundred diseases[1]. A given pathogen or pest can have a host spectrum of tens or even hundreds of plant species[2]. Plant diseases are sometimes grouped by type of symptoms (root rot, wilt, leaf spots, rust, *etc.*). However, the most useful criterion remains the classification based on the pathogen responsible for the disease, since this approach makes it possible to define the cause of the disease and the control measures to be taken. Basically, plant diseases can be classified as follows:

2.1. Parasitic (Biotic) Diseases

There are over 80,000 different parasitic diseases of plants. These diseases are caused by fungi, prokaryotes (bacteria and mollicutes), parasitic higher plants, viruses and viroids, nematodes and protozoa. Some crops have only one or two common diseases; others have many more. For example, wheat and tomato have 123 and 76 different parasitic diseases, respectively (The American Phytopathological Society website, Common Names of Plant Disease; https://www.apsnet.org/edcenter/resources/commonnames/Pages/default.aspx). Only a few of these commonly occur and cause economic problems. Quite frequently, the same plant individual may be affected by two or more diseases at one time.

2.2. Noninfectious (Abiotic) Diseases

They may be caused by:

- Temperature (too low or too high).
- Lack or excess of humidity.
- Lack or excess of light.
- Lack of oxygen.
- Atmospheric pollution.
- Nutritional deficiencies.
- Mineral toxicity.
- Soil acidity or alkalinity.
- Pesticide toxicity.
- Poor cultural practices.

3. ECONOMIC IMPACT OF PLANT DISEASES

Plant diseases reduce agricultural yield as well as the selective value of plants.

3.1. Quantitative Effect on Production

On a global scale, pathogens and pests are reducing crop yields by 10 percent to 40 percent[3]. Added to this are losses due to competing plants (weeds) and post-harvest losses. In some countries of the world, the consequences of such losses can go as far as famines. Thus, the effects of plant diseases on production can cause serious social and economic problems. The history of phytopathology is marked by many devastating epidemics, including that of 1942 when the Indian rice crop was destroyed up to 90% by *Heminthosporium orizae*, which caused a famine responsible for the death of thousands of people (the great Bengal famine) (Padmanabhan 1973).

3.2. Effect on Product Quality

Phytosanitary quality can also be affected. For example, certain mycotoxins secreted by pathogenic agents can make the products unfit for consumption or for use in processing. Even inoffensive, certain diseases can harm the appearance of products, and therefore their marketing. This is the case for certain affections of *Botrytis cinerea* causing spots on the skin of the fruit (strawberries, grapes, apples, *etc.*) without however affecting the organoleptic qualities of the latter.

4. DIAGNOSIS AND IDENTIFICATION OF DISEASES

The diagnosis of a disease, the detection of the infectious agent and its characterization are an essential prerequisite for the implementation of appropriate controlling strategies. Similarly, they make it possible to predict the upcoming crop losses, which will be caused by the infection of phytopathogenic agents (Jeong *et al.* 2014).

4.1. Diagnosis Based on Symptoms, Landscape, and Agricultural History

The diagnostic stage requires collecting different types of information.

a. Symptoms

It is necessary to define their nature (alteration of color, morphology, growth, metabolism), as well as their location (root, stem, uniform alteration of the whole plant, *etc.*). When identifying symptoms, the phytopathologist should keep in mind that symptoms affecting a specific part of the plant may have a primary organ origin.

b. The Agricultural Landscape and History

- The exposure of the plot (wind, rain, sun, *etc.*)
- The phytosanitary state of neighboring plots.
- The health status of indicator plants.
- The proximity of polluting human activities.
- Weather conditions.
- The existence of crop rotation.
- The treatments carried out (fungicides, insecticides, *etc.*)
- Origin of seeds, plants or grafts.

Symptoms may be typical when a plant is known to host a suspected pathogen, under well-defined climatic conditions and in well-defined favorable periods. In this case, the diagnosis will be easy. This is often the case with rust, powdery mildew, Smut (fungus), *etc*. However, the symptoms may be confusing and may not allow an accurate diagnosis. This is, for example, the case of certain pathogens that cause similar symptoms or symptoms varying according to the environment, or certain combined actions of abiotic and biotic stresses. These are cases where precise detection and identification of the pathogen will be required after taking samples at different stages of the disease.

4.2. Detection and Identification of Pathogens

The choice of the detection method depends on a number of criteria:

- cost (cheapest possible method).
- speed (time) and simplicity.
- risks (risk-free method of implementation).
- precision, sensitivity and reliability.

The methods can be classified as follows:

a. Methods Based on Morphological Observations

The observation can be carried out with the naked eye, with a binocular microscope, with an optical or electronic microscope. Observations can be made with or without isolation and culture.

b. Methods Based on Biochemical Markers

We can mainly cite the identification using iso-enzymatic markers (Bonde *et al.* 1993).

c. Serological/Immunological-Based Detection Systems

The serodiagnosis of phytopathogenic agents may be performed by the ***enzyme-linked immunosorbent assays (ELISAs)*** technique (Jeong *et al.* 2014). ELISA is antibody-antigen based assay, where the intensity of infection is measured by the optical density (degree of coloring) of the reaction. An important advantage of ELISAs is the fact that a large number of samples can be tested at the same time. Nevertheless, ELISAs for the detection of plant viruses have also some limitations such as the accessibility of antibody for target virus, cost of antibody production, high sample volume requirements, and time to complete ELISAs. Generally, ELISA is less sensitive than molecular methods, and misdiagnosis due to false positives[4] is likely.

d. Methods Based on Molecular Markers

Molecular markers are based on nucleic acid analysis. As an alternative method to serology, molecular markers are commonly used in the laboratory due to their high accuracy and sensitivity. We can cite the techniques PCR, real-time PCR, RAPD (Williams *et al.* 1990), RFLP (Botstein *et al.* 1980), AFLP (Vos *et al.* 1995), microsatellites, *etc.* Many of these techniques have been used to detect, characterize and map pathogen genes. Some PCR-derived techniques (*e.g.* RAPD, AFLP) don't require a specific knowledge of the pathogen genome sequence. On the other hand, PCR assays are developed to target ubiquitous, highly conserved genes or regions, such as 16S rDNA (for bacteria), or internal transcribed spacers (ITS), using universal primers. Those genes have conserved sequences and therefore lack substantial polymorphism between closely related bacterial species or strains that have distinct virulence features. Since the advent of high-throughput sequencing, hundreds of whole genome sequences from bacterial, fungal, oomycete, nematode, viral and viroid plant pathogens have been published and stored in databases (Hamilton *et al.* 2011; Zhang *et al.* 2018). The availability of such sequence data has been helpful to characterize species and strains in many studies (Thynne *et al.* 2015; Aylward *et al.* 2017; Xu and Wang 2019).

5. MOLECULAR HOST-PATHOGEN DIALOGUE

During the process of infection of a plant by a pathogen, a molecular dialogue takes place between the parasite and the plant and leads to two possible situations:

5.1. Compatible Reaction

In this case, the pathogen prevails over the plant. It actively multiplies and colonizes the host, which develops more or less accentuated symptoms depending on the degree of virulence of the pathogen and the degree of susceptibility of the plant.

5.2. Incompatible Reaction

In this case, the defense reactions of the plant are effective and stop the multiplication of the parasite. Three types of resistance can be developed by the plant:

a. Non-Host Resistance [5]

Non-host resistance often corresponds to the absence of pathogenicity rather than the presence of resistance. It is a broad spectrum resistance that is effective against entire groups of parasites, such as terpenes against insects, tannins against a wide range of parasites and phytoalexins against microorganisms. It should be noted that some parasites have developed mechanisms to neutralize, tolerate or overcome such large-scale resistance mechanisms. For example, cereal rust is a highly specialized pathogen that suppresses these broad resistances.

b. Horizontal Resistance [6]

Nonspecific or horizontal resistance implies resistance to all isolates from the pathogenic organism and is often determined in a polygenic manner.

c. Vertical Resistance [7]

Race-specific or vertical resistance involves resistance to certain isolates of pathogens and not to others and is often inherited following a simple monogenic pattern.

6. METHODS OF CONTROLLING PATHOGENS AND PESTS

Given the extent of the damage caused by pathogens and pests, the development of ever better targeted control methods has been, and still is, the subject of a great deal of agronomic research. These methods cover all approaches: agronomic (uprooting, manuring, tillage, rotation or cultural association, phytosanitary state of seeds), physical (solarization), chemical (use of more or less specific herbicides), genetic (resistant cultivar breeding) and biological (use of insects, fungi or bacteria). However, all the methods have economic, technical or sociological limits.

6.1. Phytosanitary Regulations

Phytosanitary regulations aim to monitor the phytosanitary status of crops, inform about the risk of diseases, prevent (by quarantine or prior disinfection) the introduction, into a given area, of plants or seeds infected with new pathogens, and issue certificates for the exploitation of plants, seeds or products.

6.2. Control by Cultural Practices

Cultural control uses adapted cultivation techniques to disadvantage the development of parasites by disrupting their life cycles. Examples include crop rotation over time and space, changes in soil pH, irrigation techniques and the removal of crop residues. Planting planning influences the development of diseases. Indeed, it is necessary to sow while avoiding that the period of sensitivity of the plants coincides with the massive arrival of the pathogen.

6.3. Chemical Control

Very fashionable since the First World War, it is currently showing both its extreme efficiency and its (sometimes) disastrous environmental consequences. This control method is useful if used with the ***right product***, at the ***right time***, in the ***right place*** and at the ***right dose***.

- ***The right product*** is determined by the disease to be fought. Herbicide, insecticide, fungicide, or nematocide, the first part of the name indicates the targeted category. It should be noted that insecticides and nematocides would eventually make it possible to fight against viruses by eliminating the vectors. On the other hand, there are broad-spectrum products that kill everything in the category, and narrow-spectrum products that kill a restricted number of species. For example, we can have weed killers that only attack broadleaf weeds.
- ***The right time*** is determined both by the development stage of the pathogen, which is most susceptible to the action of the product and by the environmental conditions at the time of application and the days that follow (temperature, wind, rain).
- ***The right place*** is determined by the site and the type of application of the product (contact[8], systemic[9], preventive[10]). If it is a contact product, it must meet the disease, often below the leaves. If it is a prevention product, it should constitute a protective barrier. If it is a systemic herbicide, it must not touch any other plant than the one targeted. Wind and rain are two decisive factors in product application errors. A treatment on a windy day comes to act on man by contact. A rain after treatment leaches the product in the soil or in the river. Contact with treated plants is also toxic.
- ***The right dose:*** Concentration of the phytosanitary product is often expressed in dose per hectoliters (hL) or %. However, a dose expressed in this way may give highly variable applications of active substance, mainly due to variable crop structures and planting systems. At the present time, great efforts are being made to obtain optimum efficacy from the applied product and to avoid unnecessary emission of products into the environment and residues in feed and food. The best way to achieve this is to adapt the dose rate to the area where the treatment is needed (*e.g.* crop canopy architecture).

The most frequent insecticides are based on ***Organophosphates*** and ***Pyrethroids***. However, these compounds are likely to cause serious damage to crops and to human health (Charbonnier *et al.* 2015). In this context, risk reduction strategies related to pesticides are developed. Furthermore, there is a trend towards the use of some natural substances, which have the property of acting as bio-pesticides. We distinguish :

- *Microbial pesticides*, which are made of microorganisms: entomopathogenic fungi (*e.g. Beauveria bassiana, Lecanicillium* spp., *Metarhizium* spp.), or entomopathogenic viruses (*e.g.* codling moth granulosis virus; Charmillot *et al.* 1998). Entomopathogenic nematodes are also often classified as microbial pesticides, although they are multicellular organisms.
- *Plant-based pesticides*, such as alkaloids, terpenoids, natural phenolic compounds, and other secondary chemicals. Also, certain vegetable oils, such as rapeseed oil, are known to have pesticidal properties. In addition, garlic (*Allium sativum*) has long been known to have uses in pest control for its repellent effects.

6.4. Physical Control

The physical control includes physical- and mechanical-based methods. Some methods are old (trapping, sampling, weeding, *etc.*) and others are more recent (electromagnetic radiation and pneumatic methods) (Boiteau and Vernon 2001).

6.5. Biological Control

Biological control refers to the intentional use of introduced or resident living organisms (other than disease-resistant host plants) in order to suppress the activities of plant pathogens (Pal and McSpadden Gardener 2006). Overall, we can distinguish two families of approaches:

a. The Strategy of Antagonistic Organisms

A first strategy consists in the use of hypovirulent or avirulent biocontrol strains, which compete with the virulent strain of a pathogen for the invasion of the host. For example, reducing the virulence was caused by the use of fungal viruses (family *Hypoviridae*) that induce hypovirulence in the competitive strain of *Cryphonectria parasitica*[11] (Double *et al.* 2018). Another example of application of this strategy was the use of a specific isolate of *Chryseobacterium* sp. that degrades the lectin needed by the rice blast pathogen *Magnaporthe oryzae* for spore attachment on the host leaf surface (Ikeda *et al.* 2013).

b. The Strategy of Secondary Plants

A second strategy is the use of trap plants, which attract, divert, intercept and/or confine the target insects or the pathogenic agents they vector, in order to reduce the damage caused to the main crop (Shelton and Badenes-Perez 2006). Parolin *et al*. (2012) reviewed the most common types of *secondary plants* that may be grown together with the primary crop, for biological control purposes. Such plants fall into several categories: companion, repellent, barrier, indicator, trap, insectary, and banker plants.

6.6. Genetic Resistance

It is based on the cultivation of varieties that are genetically resistant or tolerant to the considered pathogen/pest. The race-specific (or vertical) resistance is very effective but can be overcome by the (mutated) pathogen. On the other hand, the quantitative resistance (also named polygenic, general, horizontal) is partial, but slows the progression of the disease whatever the strain of pathogen, with more or less specificity.

6.7. Integrated Pest Management (IPM)

All the methods and techniques mentioned in the previous paragraphs are combined to obtain optimal results. Integrated Pest Management (IPM) refers to an integrative methodology that is an alternative to the exclusive use of a single approach for control, while taking into consideration the reduction of the impact on the environment.

According to a definition by the FAO, the IPM represents *a crop management system which, in the context of an environment and an evolution of pest populations, uses all the available control techniques, in the most coherent way possible, in order to keep the level of pests below the economic nuisance threshold* (Nandris *et al*. 1997). This rational management of pests must ensure that there is no total dependence on a single method of control and take into account the economic, social and environmental consequences of control strategies. It is easy to understand that this can only be obtained through multidisciplinary investigations to acquire knowledge on:

a. the effects of human intervention on the crop and on the dynamics of pest/pathogen populations.
b. effects of pests/pathogens on production.
c. economic profitability from the use of control methods.

The purpose of these integrated approaches is to obtain efficient and rapid management of the populations of pathogens, if possible even before the first damage is caused.

CONCLUSION

By sharing borders with other scientific disciplines such as molecular biology, genetics, biochemistry, physiology of plants and pathogens, ***phytopathology*** has allowed several recent advances such as the discovery of viroids, the understanding of defense mechanisms of plants and the discovery and use of the pathogenesis mechanism of *Agrobacterium tumefasciens* as a vector for molecular cloning. Updated knowledge of the relationships between host and pathogen is essential to reduce the economic impact of plant diseases by providing new appropriate control methods and a better protection of cultivated plants while respecting the environment.

[1] Fletcher *et al.* (2006) reported that, in the United States alone, plants are subject to attack by more than 50,000 different pathogens, primarily fungi, viruses, bacteria, and nematodes.

[2] As an example of a plant pest characterized by a broad host spectrum, we can cite the case of the whitefly, *Trialeurodes vaporariorum*, which has been found on plants from 249 genera in 84 plant families (Russel 1977).

[3] At a global scale, pathogens and pests are causing for five major food crops losses varying between 10 and 40 percent, as follows: wheat losses of 10-28%, rice losses of 25-41%, maize losses of 20-41%, potato losses of 8-21%, and soybean losses of 11-32%, according to a recent study published in Nature Ecology and Evolution (Savary *et al.* 2019).

[4] Due to non-specific reactions.

[5] Docmented in chapter 6.

[6] Docmented in chapter 9.

[7] Docmented in chapter 7.

[8] The product only works where it has been placed; it does not migrate.

[9] The product migrates and acts throughout the plant.

[10] Preventive treatments are applied before the appearance of visible symptoms, possibly even without the presence of the pest or pathogen, to constitute a barrier against infection.

[11] Agent of chestnut blight.

CHAPTER 4

Plant Immunity: An Overview

Abstract: Plants are called upon to defend themselves against a variety of organisms (bacteria, protists, fungi, insects and vertebrates) that use them as a source of nutrients. Plant immunity is the innate or induced ability of plants to detect and counteract invasive species before they can inflict serious harm. Molecules emitted by pathogens are perceived by receptors at the surface of or inside plant cells. This perception induces complex signaling cascades leading to a defensive response to pathogen attack. This chapter gives an overview of the most important defense strategies in higher plants.

Keywords: Acquired immunity, Avoidance, Coevolution, Innate immunity, Immune memory, Non-specific immunity, Programmed cell death, Resistance, Specific immunity, Tolerance, Zigzag model.

INTRODUCTION

Plants are exposed to several pathogens. In comparison to animals, plants are devoid of mobile immune cells serving to eliminate pathogens. They have, therefore, evolved alternative recognition and defense mechanisms. Actually, each plant cell has an innate immune system and the emanation of systemic signals transmitted from the place of infection leads to the generalization of defenses, on the whole-plant scale (Jones and Dangl 2006). The recognition of pathogens in plants calls for two types of perception. The first type of non-specific recognition involves molecules called general elicitors; the second, which is specific, was first described by Flor in 1955, when the concept of gene-for-gene was stated. These two recognition mechanisms constitute the two primary components of the plant immune function (Jones and Dangl 2006).

Dhia Bouktila & Yosra Habachi

1. COEVOLUTION OF PLANT DEFENSE AND PATHOGEN ATTACK MECHANISMS: THE ZIGZAG MODEL

According to the zigzag model presented by Jones and Dangl (2006) (Fig. **1**), plants have two lines of defense: a first called ***basal or non-specific defense*** and a second known as ***specific defense***, often called ***resistance***.

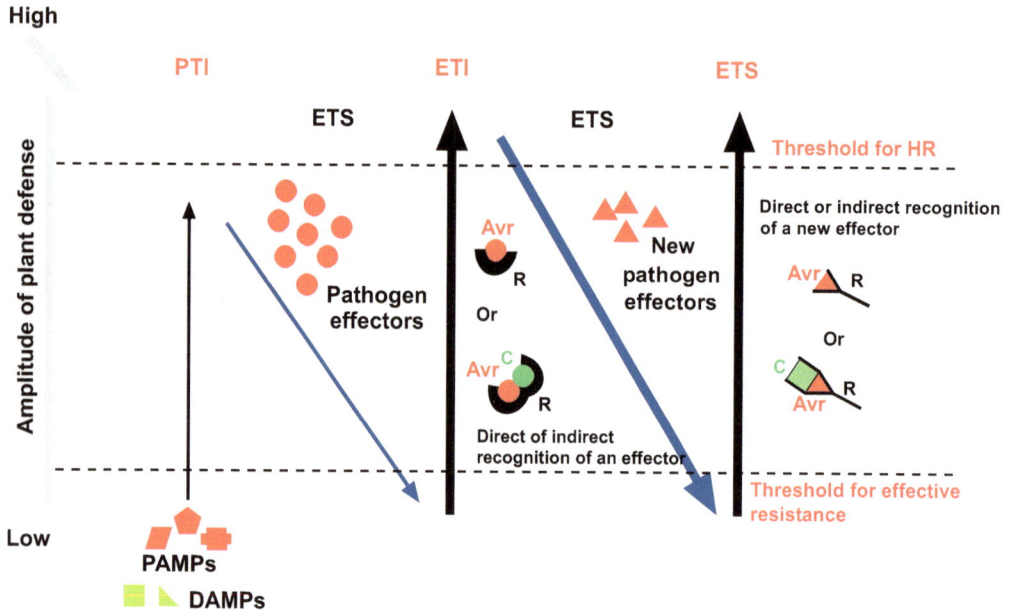

Fig. (1). Zig-zag model illustrating the innate immune system of plants and the coevolution of plant defense and pathogen attack mechanisms (modified from Jones and Dangl 2006).

The first level of defense is formed by all of the physical barriers of the cell, in particular the cuticle and the cell wall, and by non-specific defense reactions. This first basal defense (or PTI, for PAMP-Triggered Immunity) is activated when plants perceive *via* their membrane receptors (Pattern Recognition Receptor, PRRs) the pathogen-associated molecular patterns (PAMPS). Although this defese line is not specific, it makes it possible to limit the spread of pathogens. For example, the lignified cell walls represent an important barrier as they are impermeable to pathogens. In addition, some plants secrete antimicrobial molecules (such as alkaloids, phytoanticipins, phytoalexins and phenylpropanoic compounds) and generate reactive oxygen species (ROS) to limit the spread of the pathogen (Jiang and Tyler 2012).

However, a number of pathogens manage to get rid of these physical barriers by infecting plants *via* natural openings (stomata, wounds). Pathogens also produce *effectors* that favor their development while suppressing the basal defense (PTI). Therefore, the plant becomes susceptible due to these effectors. This stage of plant-pathogen interaction is called ***Effector-Triggered Susceptibility (ETS).***

Parallel to this evolution of pathogens' virulence, the plant has developed specific cytosolic receptors capable of perceiving these effectors and allowing the triggering of the second line of defense (specific defense). This specific defense system involves complex cellular events which lead to resistance to certain pathogens. Therefore, concomitant host-pathogen ***coevolution*** has enabled certain plants to develop specific resistance (*R*) genes that directly or indirectly recognize the *Avr* genes of pathogens and confer ***Effector-Triggered Immunity (ETI)*** (Jones and Dangl 2006). At this stage, the ***incompatible*** interaction of *R* and *Avr* genes activate complex signaling cascades, which can locally lead to a ***hypersensitivity reaction (HR)***, as well as to the establishment of systemic resistance at the level of whole plant.

The diagram can be described based on four co-evolutionary phases. Phase 1: Plants recognize the microbe/pathogen-associated molecular patterns (MAMP / PAMP, red patterns) using PRRs to induce PAMP-triggered immunity (PTI). Phase 2: Successful pathogens produce effectors able to overcome PTI, leading to effector-triggered susceptibility (ETS). Phase 3: An effector (full red circles: Avr) is perceived by a resistance (*R*) protein, directly or indirectly by means of a cooperating protein (C), thus activating effector-triggered immunity (ETI), an enhanced variant of PTI, frequently surpassing the induction threshold for hypersensitive reaction (HR). Phase 4: Some isolates of the pathogen gain new effectors (full red triangles: Avr), which can help them overcome the ETI. The selection will later favor plant new alleles that encode the modified resistance proteins, which are able to neutralize the newly recruited effectors. As a consequence, ETI will be once again established.

2. COMPONENTS OF PLANT IMMUNITY

Plant defense against biotic stresses has been classified as ***innate*** (either broad-spectrum or race-specific) and ***acquired*** (either local or systemic) immunity.

2.1. Innate Immunity

The plant has two forms of inherent (*i.e.* innate) defense; ***non-specific*** (*i.e.* basal or general resistance) and ***specific*** (cultivar/pathogen race specific). Therefore, the plant employs a two-level perception system: one used in non-specific defense and another used in specific defense. The first level, termed PRR-triggered immunity (PTI), is processed by surface-positioned pattern recognition receptors (PRRs). The second level, termed effector-triggered immunity (ETI), requires intracellular immune receptors, mostly belonging to the NOD-like receptor (NLR) family, which directly or indirectly target virulence effectors secreted by pathogens inside infected plant cells.

Non-specific immunity uses both ***passive barriers constitutively present*** in the plant body, and a broad variety of ***early induced*** proteins and other organic molecules that are needed early during the innate immune response and, therefore, are either formed upon infection, or preformed but released only during pathogen attack. Innate immunity in PTI is characterized by (a): Recognition of evolutionarily conserved microbe/pathogen-associated molecular patterns by receptors for microbe-associated molecules, (b): amplification of the signal through a variety of signaling cascades and (c): generation of antimicrobial peptides/compounds. Both PAMPs and DAMPs are detected by transmembrane pattern recognition receptors (PRRs). In fact, 'non-self' (PAMP) receptors include fungal chitin, bacterial flagellin, lipopolysaccharides and peptidoglycans, while specialized receptors[1] recognize the 'compromised self', called damage-associated molecular patterns (DAMPs).

One key trait of ETI is the capacity of the host to perceive a specific microbial strain and adapt to its evolutionary changes, while PTI is only able to basically recognize 'non-self' and 'infectious/compromised-self'. Qualitatively, PTI and ETI induce comparable defense responses (Fig. **2**), but ETI triggers defense mechanisms in a ***more intense and prolonged*** manner than PTI (Fig. **1**). ETI is generally extended by a localized cell death called the hypersensitive response (HR), which serves to avoid gretaer proliferation of the pathogenic attack (Fig. **3**). Resistance governed by *R* genes (ETI) is typically not durable. However, ETI is now efficiently implemented by pyramiding multiple distinct resistance (*R*) genes in the same cultivar, which tends to improve durability and resistance scope.

2.2. Acquired Resistance

One other type of resistance that has received considerable attention in plant immunity research is ***systemic acquired resistance (SAR)***, in which defense proteins are disseminated into uninfected plant parts, distantly from the initial infection area. SAR confers long-lasting protection against a wide range of pathogens and pests and contributes to a rapid and intense response during subsequent infections, based on past experience. Close to SAR, ***induced systemic resistance (ISR)*** is another type of acquired resistance that is facilitated by non-pathogenic rhizobacteria, many of which belong to the genus *Pseudomonas*.

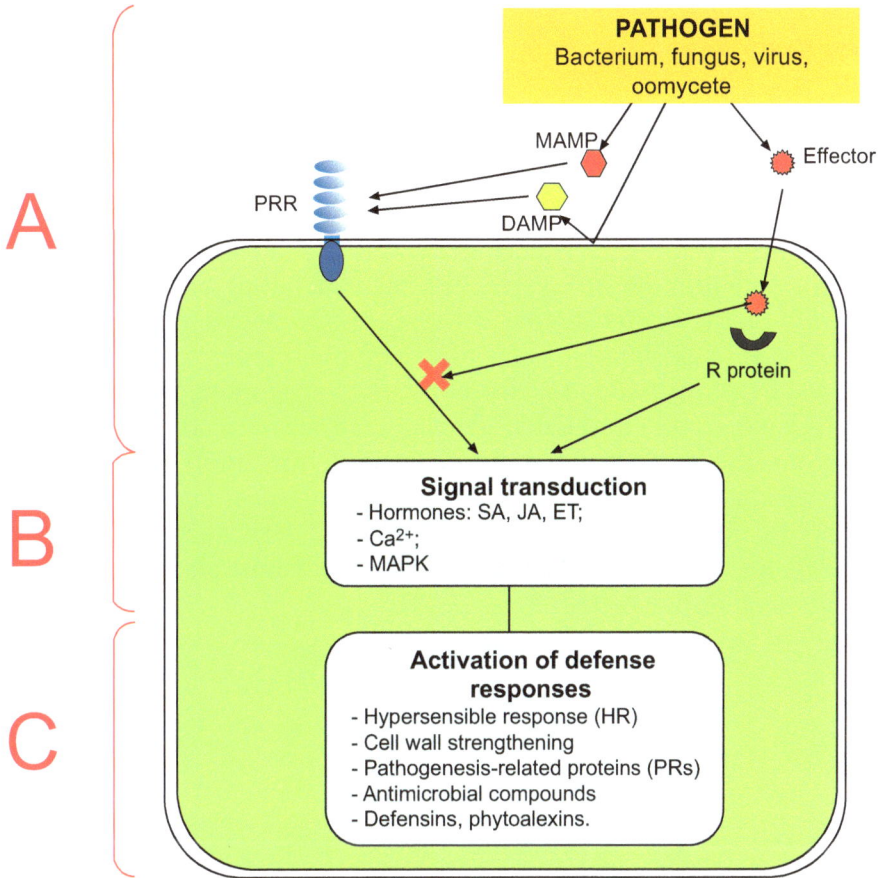

Fig. (2). Schematic representation of the innate immunity of plants.
(a) Perception of microbe-associated molecules by transmembrane or intracellular receptors: The attack of the pathogen is perceived *via* PAMPs, DAMPs (resulting from the degradation of the plant wall) or effectors. PAMP and DAMP are recognized by PRRs, inducing PTI. The effectors injected into the cell contribute to the inactivation of PTI. However, they can be recognized, directly or indirectly, by the *R* proteins, which then trigger ETI. **(b) Cascades of protein phosphorylation/dephosphorylation that may involve transcription factors regulating particular sets of stress responsive genes:** Activation of the defenses involves cellular signaling processes characterized by the release of secondary signal molecules [2] (*e.g.* Ca^{2+}, phytohormones, in particular the SA, the JA and the ET) and the activation of protein kinases (PK), including MAPKs. **(c) Stress resistance, plant adaptation and other phenotypic and physiological responses.**

2.3. Host Versus Nonhost Resistance

There are two forms of disease resistance that may be expressed by a plant, nonhost resistance and host resistance, based on the identity of the plant–pathogen pair.

a. Nonhost Resistance

Plants are by default immune to most pathogen species present in nature. This condition of the plant is called non-host resistance. The latter could be described as the resistance expressed by the whole plant species (including all its populations and members) to all strains of a given pathogen species (Nürnberger and Lipka 2005)[3]. Nonhost resistance is expressed against *unadapted pathogens* as a result of a continuum of layered defenses, including *constitutive (passive) and inducible (active) defense mechanisms.* Passive barriers are more likely to lead to nonhost resistance to parasites of distant plant species than to parasites of plant species that are taxonomically close to the nonhost (Nicks and Marcel 2009). On the other hand, inducible defense requires a broad variety of proteins and other molecules preformed (*i.e.* formed prior to infection) or produced during pathogen invasion (Onaga and Wydra 2016). There is a certain similarity between the concepts of non-specific defense and nonhost resistance, as in both cases similar molecules may be recognized by host and nonhost plants, respectively, to trigger an immune response (Gill *et al.* 2015).

Fig. (3). Hypersensitive Response on Tobacco Leaf.
Cornell Plant Pathology Herbarium Photograph Collection (http://www.plantpath.cornell.edu/PhotoLab/stills.html).

b. Host Resistance

Host resistance is known to be cultivar- or accession-specific and not very durable (Mysore and Ryu 2004). To date, most efforts made by plant breeders have been concentrated on introgressing host resistance conferred by a single or multiple resistance (*R*) genes, into susceptible cultivars. Continous monoculture[4], which is widely practiced, places significant selection pressure on pathogens, obliging them to mutate their virulence effectors in order to avoid detection by the host. As a result, the resistance of most *R* genes becomes less robust (Dangl *et al*. 2013).

3. CONCEPTS OF AVOIDANCE, RESISTANCE AND TOLERANCE

The principles of avoidance, resistance and tolerance constitute the three pillars of plant defense. ***Avoidance*** is any anti-parasite behaviour or technique exhibited by a plant, especially by at-risk hosts. Accordingly, avoidance occurs before any parasitic contact between the host and the parasite is established and it is aimed to minimize the incidence rate of the parasite on the host (*e.g.* the frequency of ovipositing). In case avoidance of the parasite fails, the host may either ***resist the parasite*** by limiting its multiplication, or ***tolerate*** the infection by reducing its fitness cost (host biomass, growth and reproduction), regardless of the level of pathogen multiplication.

Avoidance is majorly employed against animal parasites and encompasses a range of mechanisms such as volatile repellents and morphological barriers like hairs and thorns[5]. Resistance[6] usually relies on a genetic and molecular basis (Zhang *et al*. 2013). In contrast, tolerance remains largely unknown, because it is very difficult to assess and is frequently confused with quantitative forms of resistance[7]. Increased tolerance following infection or infestation is generally associated with some physiological processes and attributes that preserve or improve plant fitness following infection or infestation (*e.g.* an increased photosynthetic rate, an extensive regrowth of roots, an increased lignification that could boost the strength of the roots eaten by insect herbivores, *etc.*) (Sánchez-Sánchez and Morquecho-Contreras 2017).

4. COMPARISON BETWEEN IMMUNE SYSTEMS IN PLANTS AND ANIMALS

The identification and defense against non-self represents the core of plant and animal immunity. During (non-specific or specific) immune reactions, plant or animal receptors detect (also non-specific or specific) pathogenic elicitors, termed antigens in animal organisms. The striking similarities between innate immune systems, in plants and animals, suggest an optimized common mechanism that has evolved independently in the two kingdoms.

The identification of intrusive molecules is at the origin of the implementation of similar, but not identical, immune responses in both plant and animal organisms:

a. Non-specific Immunity (Plants *vs.* Animals)

There is a striking similarity between plants and animals in cell surface receptors, usually kinases. In PTI, there is evidence of the conservation of molecular evolution of structures and functions, beyond kingdoms borders (Nürenberger *et al.* 2004).

b. Specific Immunity (Plants *vs.* Animals)

Plants lack the antibody-providing immune system found in vertebrates. They also lack mobile immune cells similar to the lymphocytes (B and T) that can detect pathogens in the circulatory system of animals. In animals, receptor identification of pathogenic antigens causes clonal expansion and differentiation of lymphocyte receptors. In contrast, plants, which have a poorly developed circulatory system[8], use a different immune strategy to establish their specific immunity: ETI, which enables detecting specific pathogen effectors by receptor proteins encoded by the *R* genes *in each individual plant cell*. Evidence of the existence of ETI in animals is absent.

c. Immune Memory (Plants *vs.* animals)

In animals, B lymphocytes produce antigen-specific antibodies and T cells actively kill pathogen-infected cells. Both lymphocyte types defend against previously-experienced antigens and may offer lifelong immunity against certain pathogens. Plants are devoid of these memory cells. Instead, *plant immune memory* takes the form of a stress memory, which operates against *secondary infection* in distant plant organs[9].

d. Programmed Cell Death (apoptosis) (Plants *vs.* animals)

Cell death is a crucial component of innate immune responses in both plants and animals. Besides the fact that they share striking convergences and similarities in the global organization of the evolution of their innate immune system, both plants and animals may respond to infection through programmed cell death. The *hypersensitive response (HR)*, causing cell death in plants, has morphological characteristics, molecular architectures and mechanisms that recall the different types of inflammatory cell death in animals.

CONCLUSION

The various aspects of this chapter are intended to provide a general framework of plant immunity fundamentals, preparing the path for a more detailed development, in the next sections of this book, of the various facets of plant defense strategies, including the qualitative and quantitative dynamics and related molecular models.

[1] Refer to chapter 11

[2] Also called « second messengers ».

[3] It can be examplified as follows: Potato late blight pathogen does not affect an apple plant, nor does an apple pathogen, such as *Venturia inaequalis*, infect potato. In the same way, the fungus that causes powdery mildew on wheat (*Blumeria* (*Erysiphe*) *graminis* f .sp. *tritici*) does not parasitize domesticated barley and, vice versa, the fungus that causes powdery mildew on barley (*B. graminis* f. sp. *hordei*) does not infect wheat (Invasive Species Compendium, Datasheet *Blumeria graminis*, powdery mildew of grasses and cereals https://www.cabi.org/isc/datasheet/22075, last modified 24 April 2020).

[4] Continuous monoculture, or monocropping, is the agricultural practice of cultivating or raising the same crop, species, variety, or breed at a time in a farming system, year after year.

[5] These aspects will be further detailed in Chapter 5 of this book.

[6] Covered in Chapters 6-9 of this book.

[7] Addressed in Chapter 9 of this book.

[8] Plants are equipped with a passive vascular system adapted to the transportation of dissolved solutes like mineral nutrients, sugars, *etc.*, but they lack a kind of active circulation that a heart-like pump would provide.

[9] Refer to paragraph 2.2. Acquired (systemic) resistance of the current chapter. For more detailed documentation, see chapter 8 of this book.

CHAPTER 5

Passive Defenses

Abstract: Plants shield against microbes using structural defenses, antimicrobial compounds and secondary metabolites. Most plants have impermeable obstacles, such as waxy cuticles, or morphological adaptive transformations like thorns, which retard the pathogen's progression into plant tissues and prevent the release of deleterious substances such as enzymes for the degradation of walls or toxins. If a pest crosses the barriers of a plant, it is usually confronted with biochemical weapons including, but not limited to, secondary metabolites with antimicrobial ability.

Keywords: Biochemical defenses, Cuticle, Mechanical defenses, Phenolic compounds, Preformed defenses, Phytoanticipins, Wax.

INTRODUCTION

When plants are attacked by pathogens and pests, they react by active and/or passive defenses. In fact, passive immunity relies on the defenses that are constitutively exhibited by the plant body, while active immunity depends on defenses that are inducible by disease or infestation. In particular, passive defenses, which are present before infection, are used to prevent entry and spread of pathogens, and include both physical barriers and chemicals.

1. PRE-EXISTING MECHANICAL DEFENSES

The very first line of protection in the plant is a robust and impermeable major obstacle: the cuticle. First, this cuticle, consisting of ***cutin*** (a lipid substance) and ***wax*** [1], coats the surface of the aerial parts of the plant, increasing their thickness. The thickness of a cuticle varies greatly between different parts of the plant and between different plant species (Wójcicka 2015). Secondly, the ***pectocellulosic wall*** [2] is a kind of exoskeleton that envelops the plant cell plasma membrane. This wall can incorporate ***lignin***, a diverse and complex polymer of phenolic compounds that gives the cells rigidity. Lignin is a major component of wood, but the concentration of lignin may be even higher in bark than in wood (Dou *et al.* 2018). Lignified cell walls (for example of the bark) are particularly resistant to pathogens and unpleasant for insects to ingest. These two mechanical elements (cuticle and bark) safeguard plants against herbivores (including insects) and pathogens. Additional adaptative strategies against grazing vertebrates encompass leaves or branches modified into **thorns** (Fig. **1**). These organs discourage herbivores through inducing physical harm such as skin irritation, stomach pain or allergic reactions.

Fig. (1). The thorn-modified leaves in cactus plants act as a mechanical defense against predators (Image source: publicdomainpictures).

We can also cite the **mutualistic relationships** established between certain species of acacias (*Acacia* sp.) and the ants of acacia tree, *Pseudomyrmex ferruginea*; the acacias provide shelter and food to aggressive biting ants, which in turn defend the trees by attacking herbivores who try to feed on the tree leaves (Palmer *et al.* 2008) (Fig. **2**).

In addition to the structural barriers cited above, we can add to it the following (reviewed in Doughari 2015):

• **The Ctoskeleton:** Actin cytoskeleton is an important structural feature of eukaryotic cells. Particularly in plants, actin cytoskeleton stands as a major barrier experienced by pathogens in the site of infection, and its disruption was followed by cellular infiltration of many non-host fungi into the crops, as reported in wheat (*Triticum aestivum* L.), barley (*Hordeum vulgare* L.), tobacco (*Nicotiana tabacum* L.) and cucumber (*Cucumis sativus* L.) (reviewed in Doughari 2015).

- **Nectarthodes:** Hydathodes are pores present naturally in leaf ends, whose role is to excrete sugary nectar, which limits the access of sugar-intolerant microbes into the plant.

- **Leaf Hairs and Trichomes:** Leaves are covered by hairs that act as barriers to pathogen penetration. For example, in chickpea, the rich hair of the leaves and the pods represents a defensive line against *Ascachyta rabei*[3] intrusion (Doughari 2015). Trichomes are thin hairs or appendages that can grow on the surface of different parts of a plant (on roots, stems or leaves); they provide combined physical and chemical immunity especially against pest species. For example, soybean (*Glycine max*) trichomes stop the access of insect eggs into the epidermis, causing the larvae to starve after hatching (Freeman and Beattie 2008).

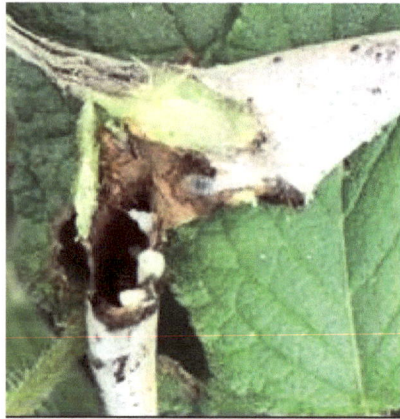

Fig. (2). Opened *Acacia cornigera* thorn, bearing adult and immature *P. ferruginea*.
(By Meixiaotian, CC BY-SA 4.0, https://commons.wikimedia.org/w/index.php? curid=76585564).

2. PRE-EXISTING BIOCHIMICAL DEFENSES

The external protection of a plant can be compromised by mechanical damage (wounds) that can constitute an entry point for pathogens. It is also possible that some micro-organisms manage to overcome the physical barriers of the plant by infecting it through natural openings such as stomata, or even through the action of hydrolytic enzymes. As soon as the primary level of plant defenses has been violated, the plant will have to use a new range of protective mechanisms based on chemical substances, such as toxins and enzymes.

Plants are rich in *secondary metabolites*, which are not necessary for the plant. Secondary metabolites are molecules that are not derived directly from photosynthesis and are not required for respiration or for the growth and development of plants. These compounds are stored in a non-toxic way in the plant and prove useful during the attack of the pathogen. When the latter manages to break the cells, the secondary compound reacts with an enzyme that deglucosylates it. An *aglycone*, toxic to the pathogen is then obtained. There are several families of secondary metabolites and other chemical barriers:

2.1. Phenolic Compounds

These are pre-infectional compounds, the content of which may increase during infection. Therefore, they participate in constitutive-type defense mechanisms. Flavonoids are one of the main groups of plant phenolic compounds. Among flavonoids, flavones and flavonols function to protect cells from harmful UV-B radiation (Lake *et al.* 2009; Saviranta *et al.* 2010). Phenolic acids can complex with each other to form *tannins*. The latter are general toxins that inhibit the growth and survival of several herbivores and serve as repellents to a number of animal species.

2.2. Terpenoids

Terpenes stand as the main group of secondary metabolites (Chen *et al.* 2018). They are composed of 5-C isopentanoid units, and behave as feeding deterrents[4] to many herbivores (Mazid *et al.* 2011). There is a wide variety of terpenes synthesized by plants, which are associated with plant immunity, acting as poisons to a vast number of plant pests and feeding mammals. Among these we can cite: (a) *Pyrethroids* (monoterpene esters) that are especially found in the flowers and leaves of *Chrysanthemum* species (family *Asteraceae*). These pyrethroids exhibit excellent insecticidal properties, due to their activity as neurotoxins to many insects such as wasps, bees, beetles, moths, *etc.* (Mazid *et al.* 2011), and (b) *Rubber* (tetraterpene), which is present in laticifers[5], offers protection to the plant as a tool for wound healing, having a defensive potential against herbivores (Surridge 2015).

2.3. Alkaloids

Alkaloids are a large group of bitter-tasting nitrogen-containing compounds present in a variety of vascular plants. As examples of alkaloids, we can cite caffeine (in coffee, *Coffea arabica*), theobromine (in tea, *Camellia sinensis*), morphine (in opium poppy, *Papaver somniferum*), nicotine (in tobacco species, *Nicotiana* sp.) and cocaine (in coca species). They are mostly made from tryptophan, aspartate, lysine and tyrosine amino acids, and all of these molecules have potent effects on animal physiology (Doughari 2015). Some alkaloids deter pests and mammals with irritating odors (like volatile mint oil) or repulsive tastes (like quinine bitterness). Other alkaloids impact herbivores by provoking either increased excitation (caffeine is one example) or the lethargy (*i.e.* fatigue and lack of energy) related to opioids.

2.4. Phytoanticipins

Anticipins are low molecular weight antimicrobial compounds present in plants before challenge by micro-organisms or are produced after infection, from ***pre-existing precursors***[6]. Phytoanticipins could be excreted outside of the plant (for example in the rhizosphere or in the leaf surface), accumulated in dead cells or stored in vacuoles in inactive state (Doughari 2015). For example, brown onions contain ***protocatechuic acid (PCA)*** sequestred inside the dead cells of their skins. Interestingly, this compound has a strong inhibitory effect on the spore germination of *Colletotrichum circinans*, an ascomycete parasite of the *Allium* family. In contrast, white onions do not stock ***protocatechuic acid*** and are sensitive to *C. circinans*. Contrary to *C. circinans, Aspergillus niger* (black mould of onion) is insensitive to PCA and, therefore, is likely to damage both brown and white onions (Devi *et al.* 2017).

Benzoxazinoids (BXs) are also another excellent illustration of phytoanticipins: BXs are commonly found across Poaceae. BX biogenesis is often subject to developmental genetic regulation, leading to the accumulation of inactive BX-glucosides that ae retained inside the vacuole. During tissue destruction, BX-glucosides are digested by plastid-targeted beta-glucosidases, leading to the production of aglycone BXs, which have biocidal properties (Morant *et al.* 2008). A number of studies have investigated the function of BXs in protecting plants against herbivorous insects and pathogens (Wouters *et al.* 2016; Niculaes *et al.* 2018).

2.5. Nutrient Deprivation

This is a form of infection avoidance, the principle of which is not supplying the pathogen with the nutritional services it needs. This may apply to both host and non-host resistance cases[7]. Spores of pathogens, such as *Spongospora subterranea* (powdery scab of potato), and eggs of the potato cyst nematode, *Globodera rostochiensis*, need special stimulants provided by the host plant to promote their germination or hatching, respectively (Guest and Brown 1980). As a result, plants that are not naturally able to produce such stimulants are immune by nature (Jibril *et al.* 2016).

It has been suggested that the non availability or limitation of nutrients to pathogens is among the strategies of non-host resistance. In other words, limiting the availability of nutrients to invasive non-host pathogens is a major reason for non-host resistance (Senthil-Kumar and Mysore 2013; Fatima and Senthil-Kumar 2015).

CONCLUSION

The passive or preformed defense mechanisms comprise mechanical barriers and strategically used bioactive chemicals that help avoid the invasion of the plant. Nevertheless, plants are not only passive hosts to the persistent invasion of microbes with which they interact. Hence, they do also develop active (induced) cellular responses to combat the occupation of their cells, once the passive approach has become obsolete. These active defense mechanisms require the functioning of the host metabolism and will be covered in the next chapters.

[1] Structurally, waxes are made of combinations of long-chain compounds of hydrophobic nature. Functionnally, they serve as water repellents on plant surface. This is critical to minimize many pathogens spore germination (Wójcicka 2015).

[2] Structurally, the cell wall is composed of cellulose, a sophisticated polysaccharide made up of thousands of glucose units bound with others, to build long polymer chains that simultaneously make the wall tough and adaptable.

[3] A necrotrophic fungus that causes a devastating disease in chickpea called *Ascochyta* blight. The disease is characterized by lesions on leaves, stems and pods (Knights and Hobson 2016).

[4] A secondary plant compound that does not kill pests, but prevents them from following food intake after initial tasting. Typically, this causes pests to die of starvation.

[5] Laticifers are highly specialized, tube-like secretory cells present in the leaves and/or stems of numerous phylogenetically unrelated (over 900 genera) plants that produce latex (a milky sap) and rubber as secondary metabolites, such as *Periploca angustifolia* (Samanani and Facchini 2006; Dghim *et al*. 2015).

[6] Phytoanticipins do not constitute a chemically unified group. Some of the examples we are citing, for phytoanticipins, correspond to phenolic compounds (it is the case of quinones catechol and protocatechuic acid, chlorogenic acid, phloridzin, arbutin and iso-chlorogenic acid).

[7] Refer to the concept of Host *versus* nonhost resistance, addressed in Chapter 4.

CHAPTER 6

Basal or Nonspecific Plant Defense

Abstract: Non-specific defense against plant pathogens can be passive (constitutive) or active (induced by microbes). The activation of general resistance follows the perception of the pathogenic threat. The first class of plant receptors recognizes molecular patterns associated with pathogens / microbes (PAMPs / MAMPs) in a non-specific way. These are resident membrane receptors, also called pattern recognition receptors, PRRs. Plant PRRs are the source of extremely complex molecular signaling immune machinery. A transmembrane receptor that binds to a ligand then triggers the signalling would be the most simplistic scenario. Yet, in many cases, the recognition scheme would also include co-receptors, as well as regulatory proteins, which activate PRRs leading to the signal trasduction intiation. It is, therefore, reasonable that our current knowledge is only touching the surface of a remarkably intricate immune strategy.

Keywords: EF-TU Receptor (EFR), Elongation factor Tu (EF-Tu), Elicitor, Flg22, FLS2, Pathogen Associated Molecular Pattern (PAMP), Pattern recognition receptor (PRR), PAMP-triggered immunity (PTI).

INTRODUCTION

The plant defensive arsenal includes a broad variety of constitutive defenses. Besides these, a sophisticated system of responses is induced upon infection that is based on the capability of plants to recognize and identify the invader.

Acting according to the wisdom prevention is better than cure, plants, first, develop protective barriers against infection. These ***constitutive, passive defenses*** are either mechanical or chemical in nature[1]. These barriers and molecules practice regardless of the presence of a well-defined pathogen, and are, therefore, considered non-specific. Their role is to protect the plant from invasion by any eventual biotic agent, by providing it strength, rigidity and vigilance.

The ***inducible, active defenses***, such as PAMP-triggered immunity (PTI), are adopted by the plant as a second resort, due to the high energy costs and metabolism mobilization, essential to their set up, activation and maintenance. However, if the pathogen has the necessary weapons to invade a plant passive defense arsenal, or the plant does not detect it, then the outcome of the interaction can be fatal for the host plant, which, therefore, will be forced to make the necessary energy expenditure. Ultimately, the plant will manage to confine the aggressor to the site of attack thanks to the establishment of active defenses triggered by its presence.

1. PASSIVE (CONSTITUTIVE) DEFENSES

The first barriers that pathogenic organisms must overcome are the physical and chemical barriers of the plant[2]. This immunity is qualified as passive because it does not imply recognition of the pathogen by the plant. The cuticle, made up of cutin (a lipid substance) and waxes, protects the surface of the aerial parts of the plant (Fig. **1**). The pectocellulosic wall envelopes each plant cell. This wall may contain lignin, a heterogeneous polymer of phenolic compounds which gives cells their rigidity. The lignified cell walls are notably impermeable to pathogens and difficult to consume by herbivores, including insects.

Some plants add antimicrobial chemical molecules (these are called secondary metabolites) to these physical barriers, such as:

a. Alkaloids, for example ***taxine alkaloids*** produced by ***yew plants***, *Taxus* spp. [3] (Wilson *et al.* 2001),

b. Terpenoids, such as digitoxin from digitalis, *Digitalis purpurea* and *Digitalis lanata* (Jackson Seukep *et al.* 2014),

c. Phenolic compounds are derived from the phenylpropanoid pathway (Kundu and Vadassery 2019), and are documented to exert a role against both insect and mammalian herbivores (Boeckler *et al.* 2011). Some examples include:

   ○ ***Chlorogenic Acid*** (3-caffeoyl quinic acid) is a derivative of caffeic acid and quinic acid (Kundu and Vadassery 2019), which provides defense against diverse insect herbivores, including the tomato fruitworm, *Heliothis zea* Boddie (Elliger *et al.* 1981) and beet armyworm, *Spodoptera exigua* Hübner (Shapiro *et al.* 2009). Consequent to wounding due to feeding by the insect, polyphenol oxidase (PPO), which is present in the chloroplast, interacts with a vacuolar phenolic substrate[4]. This interaction results in the oxidation of the substrates, which catalyzes the conversion of chlorogenic acid to an insect-toxic substance, chlorogenoquinone (CGQ).

○ *Glycoside Phloridzin* and its aglycone, phloretin, inhibit scab fungus *Venturia inaequalis*, in apple (Jha *et al*. 2009).

All of these substances are effective against a large number of pathogens. In most cases, these preexisting passive defenses constitute a sufficient obstacle against most pathogens[5].

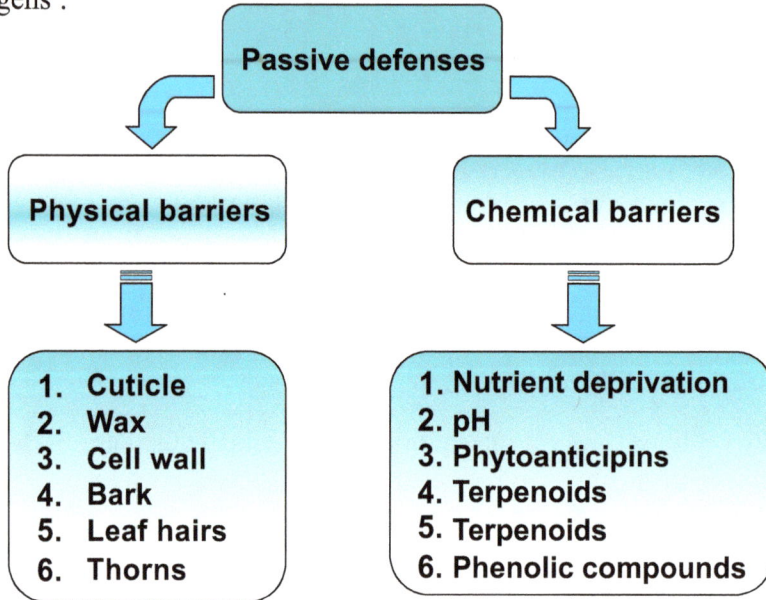

Passive defenses

Physical barriers

Chemical barriers

1. Cuticle	1. Nutrient deprivation
2. Wax	2. pH
3. Cell wall	3. Phytoanticipins
4. Bark	4. Terpenoids
5. Leaf hairs	5. Terpenoids
6. Thorns	6. Phenolic compounds

Fig. (1). The constitutive, physical and chemical, barriers of the plant.

2. ACTIVE (INDUCIBLE) DEFENSES

Some microorganisms manage to overcome passive defenses, by infecting the plant through natural openings such as stomata, through injuries, or even directly, through mechanical force and enzymatic softening of the cell wall substances[6] (Fig. **2**). These molecules can either degrade the cuticle and the cell wall, or neutralize toxic molecules by metabolizing them.

Plasma membrane-localized receptors play a major role in detecting pathogen-associated molecular patterns (PAMP) or endogenous signals released after attack, so called danger-associated molecular patterns (DAMP). Recognition of PAMPs by the appropriate pattern recognition receptors (PRR) in plants gives rise to PAMP-triggered immunity (PTI), a basal immune function active against a wide range of microbes.

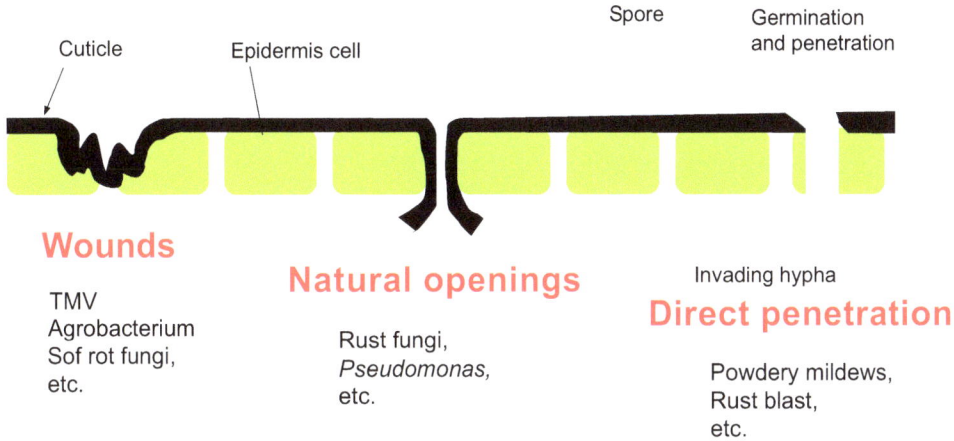

Fig. (2). The methods of penetration of infectious agents within the plant.

2.1. Development of the Concept of PAMP from that of Elicitors

The pathogen's signal molecules (or elicitors) are capable of triggering a response from the plant, which, in turn, will synthesize metabolites allowing it to defend itself. Many elicitors are *oligosaccharides*. Among this large and varied molecular family, only some have been characterized in terms of structure and biological activity. Initially, the term elicitor was used to describe the molecules capable of inducing the production of phytoalexins (Keen 1975), then it was extended to all of the molecules that induce defense reactions in plants (Montesano *et al*. 2003). The notion of elicitor then evolved to that of PAMP (*Pathogen-Associated Molecular Pattern*) due in particular to convergence of results between the work carried out on the innate immunity of mammals and plants (Medzhitov and Janeway 1997; Nürnberger and Brunner 2002). A PAMP is a molecule that functions for the host plant as a signature for the presence of any pathogen of a given class. The concept of PAMP has been further broadened to include the following types:

- PAMPs : *Pathogen-Associated Molecular Patterns*
- MAMPs : *Microbe-Associated Molecular Patterns*
- HAMPs : *Herbivore-Associated Molecular Patterns*
- NAMPs : *Nematode-Associated Molecular Patterns*
- and VAMPs : *Virus-Associated Molecular Patterns*

Another category is added, namely signals associated with self-derived damage, called DAMPs: ***Danger-Associated Molecular Patterns***. The latter are degradation products derived from the plant following its invasion by a pathogen.

2.2. Generic and Conserved Nature of PAMPs

PAMPs refer to small, well conserved (*i.e.* slow evolving) molecules secreted on the plant surface by a given class of microbes. The structures playing the role of PAMPs are conserved within various classes of microbes, where they are usually linked to ancestral functions (Nürnberger and Brunner 2002). Thus, the genetic factors encoding PAMPs are unlikely to be altered or lost during the evolutionary history of microbes. The generic interaction of plants with these conserved molecules confers to plants broad-spectrum resistance to groups of microbes displaying the same PAMP. Even with this conservation, certain PAMPs may be submitted to selective pressures during the co-evolution of the pathogen with its host plant(s). Hence, modifying some important amino acids will be helpful to the adapted pathogens to resist detection by the host plant (Boller and Felix 2009; Monaghan and Zipfel 2012).

In bacteria, the best-described PAMPs include: i) **flagellin**, a key protein crucial to bacterial mobility (Ramos *et al.* 2004), ii) **peptidoglycan (PGN)**, a cell wall polymer common to all bacteria (Janeway and Medzhitov 2002), and iii) **lipopolysaccharides (LPS)**, a cell wall element present in Gram negative bacteria. In fungi, the most represented PAMPs are **chitins** and **mannans** (Cohn *et al.* 2001; Parker 2003).

2.3. Pattern Recognition Receptors (PRRs)

PAMP-triggered immunity (PTI) plays a vital role in all higher plants (Boller and He 2009). Plant cells have a substantial number of cell-surface-membrane mounted receptors that are essential for detecting external signals. These Pattern-Recognition Receptors (PRRs) behave, therefore, like *cellular antennas*, allowing plants to perceive numerous and varied threat signals, associated with *non-self* (PAMPs, HAMPs, NAMPs and VAMPs) or self-derived damage (DAMPs).

In comparison to mammals, which use both cell surface and intracellular PRRs to sense PAMPs and DAMPs, all known plant PRRs have been found to be located on the cell surface. PRRs can be from the Receptor-Like Kinase (RLK) or Receptor-Like Protein (RLP) families (Tör *et al.* 2009). RLKs consist of an extracellular domain (ECD) also called ectodomain involved in ligand binding, a single-pass transmembrane (TM) domain and an intracellular kinase domain. On the other hand, the structure of RLPs is distinguished from that of RLKs by a unique feature: the RLPs lack of an intracellular kinase domain (Jamieson *et al.* 2018).

Plant genomes encode a multitude of RLPs and RLKs that have the potential to operate as PRRs. Yet, for many of these PRRs, the precise nature of corresponding ligands (and therefore, the precise nature of the formed immune complexes) is still uncertain. However, a number of structural, biochemical and genetic studies have successfuly explained the molecular mechanisms of PRR binding to their corresponding ligand, which could be either a PAMP or a DAMP (Wu and Zhou 2013). This molecular mechanism depends, to a large extent, on the nature of the ectodomain present in the RLP or RLK, as it represents the functional region positioned outside the cell, which initiates the interaction with the ligand. It has been demonstrated that PRR ectodomains (*i.e.* extracellular domains) may comprise leucine-rich repeats (LRRs), lectin motifs [which include Lysine motif (LysMs) and malectins], PR-5 family (Pathogenesis-related protein 5), or epidermal growth factor (EGF)-like domains[7] (Wu and Zhou 2013; Restrepo-Montoya *et al.* 2020). Within RLKs alone, the classification based on the extracellular domains defines at least 15 subfamilies (Galindo-Trigo *et al.* 2016). In particular, extracellular LRR domains present in many RLKs and RLPs adhere to pathogen-derived peptides or proteins, such as bacterial flagellin, endogenous Pep peptides or bacterial EF-Tu (Chinchilla *et al.* 2006; Yamaguchi *et al.* 2006; Wu and Zhou 2013). On the other hand, RLKs and RLPs carrying other ectodomains mainly identify carbohydrate-containing molecules, including bacterial peptidoglycans, fungal chitin, or plant-cell-wall-derived oligogalacturonides (Miya *et al.* 2007; Brutus *et al.* 2010; Choi *et al.* 2014).

2.4. Popular Models of PTI in Plants

2.4.1. Flagellin-Induced Resistance

Bacterial flagella are often detected by plants throughout basal resistance. Flagellin is the core component of bacterial flagellum, and is, one of the best-studied PAMPs in plants. A 22 aminoacid epitope (flg22) located in the N-terminal portion of *Pseudomonas syringae* flagellin is capable of stimulating the immune system of a wide range of plant species, which will consequently develop an immune reaction (Felix *et al.* 1999). In the model angiosperm *Arabidopsis thaliana*, flagellin detection is provided by a leucine-rich repeat-receptor-like kinase (LRR-RLK) named FLAGELLIN-SENSING 2 (FLS2). The ectodomain of *A. thaliana* FLS2 (At-FLS2) is made of 28 tandemly organized LRRs. These LRRs bind to flg22, causing, quasi-immediately, the formation of a

heterodimer (or a receptor complex) with another regulatory LRR-RLK, BRI1-associated kinase 1 (BAK1), that plays a role of co-receptor or adapter (Fig. **3**).

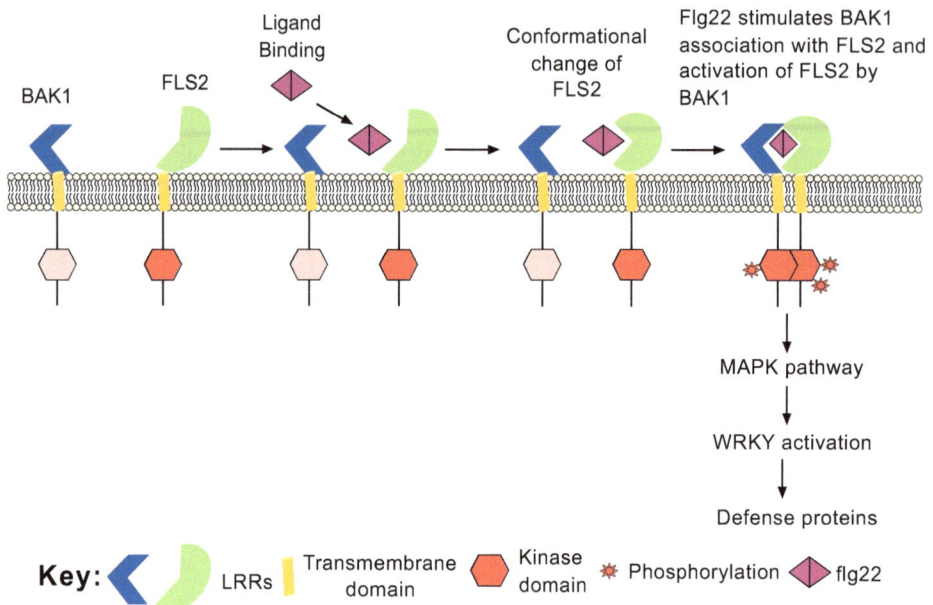

Fig. (3). Bacterial PAMP flagellin (flg22) perception by *Arabidopsis* FLS2, mediated by heterodimerization involving FLS2 and BAK1, and activation of the subsequent MAPK cascade, WRKY transcription factors and defense proteins (modified from Chinchilla *et al.* 2009).

a. Conservation, Diversity, Evolution

Orthologs[8] of *A. thaliana* FLS2 have been identified in other imprtant angiosperm species, including dicotyledons such as tomato (*Solanum lypersicum* L.), grapevine (*Vitis vinifera* L.), or tobacco (*Nicotiana benthamiana* Domin.) and monocotyledons such as rice (*Oryza sativa* L.). The existence of these homologs strongly indicates that the flagellin perception system is characterized by an ***ancient origin and evolutionary conservation*** (Felix *et al.* 1999; Boller and Felix 2009). These ***diversified FLS2 orthologs*** display distinct perception characteristics, associating them to the molecular motifs of flagellins from different phytopathogenic bacteria (Chinchilla *et al.* 2006; Robatzek *et al.* 2007; Takai *et al.* 2008; Trdá *et al.* 2014). This indicates that the flagellin receptor FLS2 is likely to experience ***functional divergence***, and that flagellin perception is an ***adaptive process*** in different plant species, resulting in diverse recognition specificities.

In most cases, this functional variability is due to mutational differences within the structural models of the extracellular LRR domains for diverse FLS2 proteins. In fact, mutation changes affecting a limited numbers of LRRs usually modify flagellin binding (Dunning *et al.* 2007). However, some mutations occurring within the intracellular kinase domain of FLS2 have been also found to strongly reduce flg22 responsiveness (Gomez-Gomez *et al.* 2001).

An interesting illustration of the selection pressure triggering FLS2 functional divergence has been reported by Takeuchi *et al.* (2003). In fact, the epigenetic, posttranslational glycosylation state (*i.e.* glycosylated *vs.* nonglycosylated or slightly glycosylated) of the flagellins of different strains of *Pseudomonas syringae* were revealed as a determining factor for the occurrence of the recognition event between a plant species and its adapted phytopathogenic bacteria.

The receptor kinase flagellin sensing 2 (FLS2) is a ***basal (or generalist)immunity*** component responsible of flagellin perception, as a part of the perception of the ***conserved*** pathogen-associated molecular patterns (PAMPs). Although functional differences and perception specificities have been demonstrated and reported, as expained above, still ***the evolution of this system of FLS2 is fundamentally different from that characterizing R-genes coding Pathogen race-specific resistance proteins***[9]. Actually, as concluded by the study of Vetter *et al.* (2012), ***PAMP detection evolves mainly quantitatively, while the evolution of R genes is mainly qualitative and operates through punctual changes in the specificity of recognition***[10]. In other words, in PAMP receptors evolution, such as FLS2, evolutionary differences between species mainly concern the receptor ***protein concentration***; and the differences observed correlate with both the ***intensity*** of the defense reaction and the spread of bacteria. From a genetic viewpoint, Vetter *et al.* (2012) demonstrated that such an evolution pattern is not only triggered by the evolution of the receptor locus, but also an evolution occurring within components common to pathways downstream of PAMP perception, which likely contributes to the quantitative variability observed within as well as between related plant species.

b. Downstream Signaling Cascade

Signaling events caused in the plant cell, following flg22 binding, start with the binding of FLS2 to BAK1 (BRI1-associated kinase 1) (Fig. **3**). Following this, the plasma membrane-located receptor-like cytoplasmic kinase BOTRYTIS-INDUCED KINASE 1 (BIK1)[11] associates with the complex FLS2/BAK1. BIK1 is initially phosphorylated upon flagellin detection, and then transphosphorylates FLS2/BAK1 to spread defense signaling (Lu *et al.* 2010). The signalling pathway downstream of flg22 perception involves the Ca^{2+} burst, the production of reactive oxygen species (ROS) and the subsequent activation of MAPK cascades. These signal transduction events invoke transcriptional reprogrammers such as the WRKY transcription factors (TFs)[12], which is crucial for the coordinated functioning of defense gene networks (Onaga and Wydra 2016).

2.4.2. Elongation Factor (Ef-Tu)-Induced Basal Resistance

EF-Tu protein is a prokaryotic elongation factor that plays a key role in gene translation by associating with amino-acyl tRNAs that supply amino acids to the ribosome. It is one of the most abundant prokaryotic proteins, representing ~5% of the total cell protein (Parker 2001). It was initially isolated from *Escherichia coli*, and acts as PAMP in members of the *Brassicaceae* family, including *A. thaliana* (Kunze *et al.* 2004). Similar to flg22 in flagellin, a synthetic peptide, elf18, corresponding to the conserved N-acetylated epitope (first 18 amino acids of the protein) is sufficient for recognition and activation of immune reactions in plants (Zipfel *et al.* 2006). EF-Tu is detected by the LRR-RLK EF-TU RECEPTOR (EFR) belonging to the same subfamily (XII) as FLS2 (Zipfel *et al.* 2006).

Similar to FLS2, the resistance function of EFR is driven by the creation of a heteromeric complex involving the BAK1 co-receptor. In *Arabidopsis*, the EF-Tu receptor EFR and BAK1 co-receptor have been demonstrated to come into physical contact, once the ligand is recognized (Kemmerling *et al.* 2011). Subsequently, downstream signal transduction is triggered by the rapid phosphorylation of EFR. The RLK BIK1 plays a key role in the transmission of signals, in a similar way to flagellin-induced resistance, as described above[13]. Downstream signalling pathway typically requires the activation of a RING finger ubiquitin ligase (XB3), XA21-binding protein XB25, MAPKs, and WRKY TFs (Onaga and Wydra 2016).

3. HETEROLOGOUS EXPRESSION OF PRR GENES

Heterologous expression of pattern recognition receptors (PRRs) is a growing research topic that aims to expand the range of crop species that apparently lack homologous PRR receptors, similar to the *Arabidopsis thaliana* EFR (*At*-EFR) and FLS2 (*At*-FLS2).

Although the potential to interact with elf18 epitope is limited to the *Brassicaceae* plant family, heterologous interfamily expression of EFR in the *Solanaceae*, more specifically in tomato (*S. lycopersicum*) and tobacco (*N. benthamiana*), conferred responsiveness to elongation factor Tu from different genera of phytopathogenic bacteria (Lacombe *et al.* 2010). More recently, *At*-EFR was also used in potato (*S. tuberosum*) (Boschi *et al.* 2017) and wheat where it triggered immunity against *Pseudomonas syringae* pv. *oryzae* (Schoonbeek *et al.* 2015). Rice plants transformed with *At*-EFR showed to be more resistant to the elf18 peptide derived from *Xanthomonas oryzae* (Lu *et al.* 2015). All these reports strongly suggest that EFR is likely to be employed for establishing large-spectrum disease resistance spanning various plant orders and families, outside *Brassicaceae*.

As for the heterologous expression of FLS2, Hao *et al.* (2016) reported that transgenic *Citrus* expressing the FLS2 receptor from *Nicotiana benthamiana* showed lower susceptibility to *Xanthomonas citri*.

Furthermore, intergeneric transfer of the PRR *Xa21* from rice (to which it confers resistance to *Xanthomonas oryzae* pv. *oryzae*, *Xoo*) to another monocotyledonous plant, namely banana, provided complete resistance to the banana Xanthomonas wilt, *Xanthomonas campestris* pv. *musacearum* (Tripathi *et al.* 2014).

Another example of successful inter-species gene transfer is the tomato RLP, *Ve1*, which detects *Ave1* from *Verticillium dahliae* race 1. This gene provided permanent resistance when expressed as a foreign gene in *Arabidopsis* (Fradin *et al.* 2011).

The number of PRR genes that have been used to improve plant resistance to bacterial pathogens through transgenic approaches remains relatively limited, to date. Because the aim of improving agricultural plants, *via* heterologous expression of wide-spectrum disease resistance, is to engineer crops that express better resistance, more PRRs that recognize conserved PAMPs will have to be characterized and their mechanism of action, implication in the physiology and metabolism of the plant and interaction with the environment, addressed.

CONCLUSION

Plant immune system relies heavily on plasma membrane-located immunity receptors that detect Pathogen-Associated Molecular Patterns (PAMPs). The identification of PAMPs by the Pattern Recognition Receptors (PRRs) represents an initial signal, which will be amplified through the activation of many downstream molecules. The cellular response insured by signal transduction cascades involve the alteration of the expression (*i.e.* overexpression or underexpression) of a repertoire of key genes, ultimately leading to plant resistance. It is nevertheless noteworthy that, if this first defense strategy is overcome, the plant will activate a second, more sophisticated, race-specific defense mechanism termed Effector-Triggered Immunity (ETI).

[1] Review Chapter 5.

[2] Refer to the previous chapter (5) for a detailed analysis of plant constitutive physicochemical defences.

[3] A conifer that may be known as common yew, English yew, or European yew.

[4] Catechin, epicatechin and caffeic acid are among the most famous PPO substrates.

[5] Generally those that are not considered as highly aggressive.

[6] Named cell wall–degrading enzymes (CWDEs) or plant cell wall–degrading enzymes (PCWDEs).

[7] EGF-like repeats are present in an important RLK family: the Cell-wal--associated kinases (the WAK family) (Restrepo-Montoya *et al.* 2020).

[8] Orthologs are homologs from different species evolved from a common ancestor.

[9] Refer to chapters 7, 10 and 12.

[10] In specific resistance, *R*-gene evolution is based on the evolution at the receptor locus alone, as a consequence of the selection pressure exerted by the mutation occurring in the corresponding pathogen effector.

[11] BIK1 performs a key role in the transmission of cellular signals not only from FLS2, but also from other PRRs such as EFR, CERK1 and the DAMP (Pep1) receptor kinases PEPR1/PEPR2 (Liu *et al.* 2013).

[12] Transcriptional reprogramming refers to the occurrence of massive changes in gene expression, where some particular genes are induced (expressed or overexpressed), and others repressed (underexpressed or silenced). This situation is usually caused by transcription factors, such as the WRKYs. The occurrence of this phenomenon in relation to plant defense is outlined in Chapter 16 of the present book.

[13] See **2.4.1.** Flagellin-induced resistance. b) Downstream signaling cascade.

CHAPTER 7

Pathogen Race-Specific Resistance

Abstract: As part of a host-pathogen coevolution, plants have developed very specific resistance proteins (*R* proteins), which directly or indirectly recognize *Avr* proteins and thus activate host resistance, or effector-triggered immunity (ETI). ETI can be thought of as an enhanced version of PTI and is often described as leading to localized programmed cell death in infected tissue: the hypersensitive response (HR). This chapter aims to clarify what we know and to identify areas that require further investigation.

Keywords: Effector-triggered Immunity (ETI), Effector recognition, Gene-for Gene resistance, Host cultivar, Hypersensible reaction (HR), Nucleotide Binding Site domain (NBS), Pathogen race, Pathogen effector.

INTRODUCTION

In specific resistance, the ***strains of the pathogen*** are capable of infecting all or part of the genotypes of the host plant species. Therefore, this is a second level of plant recognition, when the pathogen is able to cross the barrier of nonspecific recognition.

The initial model, labeled ***gene-for-gene***, assumed the existence of a physical interaction between an *Avr* elicitor and an *R* protein. This model was developed in the 1940s by the American biologist Harold Flor who studied the interactions of flax and the flaxseed rust pathogen, *Melampsora lini*. This hypothesis states that for each resistance (*R*) gene in a plant, there is a corresponding avirulence (*Avr*) gene in the pathogen. The interaction between the two corresponding genes, leads to incompatibility (resistance). This model hypothesizes that there could be a direct or indirect physical interaction between a ligand produced by the pathogen and a corresponding plant receptor, which ultimately triggers the activation of defense genes downstream.

This hypothetical model was later confirmed by molecular evidence of direct contact between the two types of proteins. Indeed, Martin *et al*. (1993) provided first evidence of direct interaction of tomato *Pto* gene with *avrPto* from *Pseudomonas syringae* pv. tomato. Later, additional evidence on this model was

Dhia Bouktila & Yosra Habachi

provided, such as ***Pita/AvrPita*** in rice (Jia *et al.* 2000), ***RRS-1/PopP2*** in *A. thaliana* (Deslandes *et al.* 2002) [1], and ***L/AprL567*** in flax (Dodds *et al.* 2006; Wang *et al.* 2007) [2]. However, cases of direct R-Avr interaction remain rare, suggesting a more complex relationship (Ellis *et al.* 2000).

It has been demonstrated that the establishment of resistance in the plant requires the presence of a gene in each of the two partners: in the plant, a gene called resistance (*R*) gene, and in the pathogenic agent, an avirulence (*Avr*) gene. In the absence of one of the players, the disease develops. On the basis of this genetic model called gene-for-gene resistance, an interpretation of the model has been proposed according to which the *R* proteins act as receptors which specifically and directly or indirectly [3] bind to a corresponding Avr ligand protein, in order to activate the plant's defense mechanisms (Gabriel and Rolfe 1990).

1. THE FLOR MODEL

Harold Henry Flor's hypothesis corresponds to an explanatory model proposed in the middle of the 20th century on the flax/flaxseed rust pathosystem. *For each gene conditioning resistance in the host plant, there is a specific complementary gene conditioning pathogenicity in the pathogen.* In other words, a plant carrying a given *R* gene will only be resistant to a strain of pathogen carrying the corresponding (specific for this *R* gene) *Avr* gene. This model has been shown to be extremely fertile and remains the basis of phytopathology today. Today, ***avirulence proteins are considered to be protein effectors***. These are proteins from the pathogen that are potentially involved in suppressing the host plant's defense reactions. When these are recognized by resistance proteins from the plant cell, the reactions triggered from this recognition lead to an incompatibility reaction.

2. PATHOGEN EFFECTORS

The notion of effector refers to any secreted molecule associated with an organism, which alters the physiology, structure or function of another organism. In particular, effectors are pathogenic molecules which can modify or even suppress the defense mechanisms induced by PTI, to thereby facilitate access to nutrients, proliferation and growth of the pathogen (Göhre and Robatzek 2008).

The effectors mainly refer to small secreted proteins, which are rich in cysteine and do not have a clear homology with other known proteins (Göhre and Robatzek 2008). Secreted effectors attain their cellular target either at the intercellular interface of the host and the pathogen cells (***apo effectors***) or inside the host cells (***cytoplasmic effectors***) (Kamoun 2006; Schornack *et al.* 2009; Djamei *et al.* 2011) (Fig. **1**).

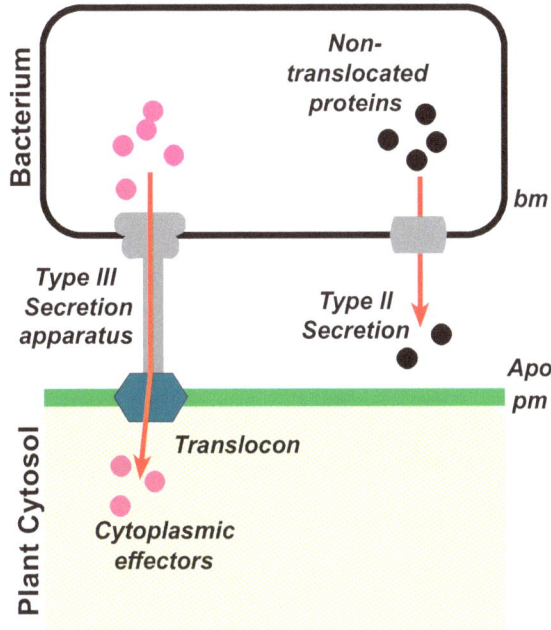

Fig. (1). Schematic view of the two classes of effectors, apoplastic and cytoplasmic, secreted by phytopathogenic bacteria (modified from Schornack *et al.* 2009). Apoplastic effectors (black circles) are secreted in the intercellular space using type II secretion system. Once in the apoplast, they interfere with the apo defenses of plants. Cytoplasmic effectors (magenta circles), on the other hand, are translocated inside the cytoplasm of the host cell. They must cross both the membranes of the pathogen and the plant. Host translocation is achieved by the type III secretion apparatus. Plant plasma membrane (pm), bacterial plasma membrane (bm), apoplast (apo).

Although the presence of effectors has been inferred for many years and their activity has been hypothetically linked to the gene-for-gene hypothesis, the application of molecular techniques, in last decades, has allowed the characterisation of a few gene-specific elicitors (*i.e.* effectors). Since the 1980s', a series of race-specific peptide products of the avirulence genes of *Fulvia fulva*, a biotrophic pathogen of tomato, has been identified. These peptides were first isolated from intercellular fluids of infected leaves and have since been found around the infection site (De Wit *et al.* 1986; Schottens-Toma and DeWit 1988). Recent research is, now, beginning to reveal the function of increasing numbers of fungal effectors bringing forward new technologies that improve our understanding in plant-pathogen interactions (Table **1**).

Table 1. Some effectors of plant pathogenic fungi and oomycetes that have been cloned and studied (extensively reviewed in Selin *et al.* 2016). [4]

Effector Protein	Length (*amino acid* residues)	Localization in the Plant	R-protein association; Role in Virulence
Blumeria graminis f. sp. *hordei* (biotroph; host: barley)			
AvrA10	286	Cytoplasm	Interaction with *Mla10*
AvrK1	177	Cytoplasm	Interaction with *Mlk1*
Cladosporium fulvum (syn. *Passalora fulva*) (biotroph; host: tomato)			
Avr 2	78	Apoplast	Interaction with *Cf2*. Binds and inhibits tomato cysteine proteases Rcr3 and Pip3
Avr4	135	Apoplast	Interaction with *Cf4*. Protects against chitinases
Avr4E	121	Apoplast	Interaction with *Cf4-E*
Avr9	63	Apoplast	Interaction with *Cf9*
Fusarium oxysporum f. sp. *lycopersici* (hemibiotroph; host: tomato)			
Avr1 (Six4)	242	Xylem	Interaction with *I-1*
Avr2 (Six3)	163	Xylem (translocated to cytoplasm)	Interaction with *I-2*
Avr3 (Six1)	284	Xylem	Interaction with *I-3*. Aggressiveness determinant.
Magnaporthe oryzae (hemibiotroph; host: rice)			
Avr-Pita 1	224	Cytoplasm	Interaction with *Pi-ta*
PWL1	147	Cytoplasm	Unknown
PWL2	145	Cytoplasm	Unknown
ACE1	4035	Fungal Appresorium	Encodes a polyketide synthase; Insertion of MINE retrotransposon into ACE1 gene responsible for virulence in strain 2/0/3
Avr-CO39	89	Cytoplasm	Interaction with *CO39*; Interaction with RGA4/RGA5
AvrPiz-t	108	Unknown	Interaction with *Piz-t*. Reduces *flg-22* and chitin induced ROS generation; Targets U3 ubiquitin ligase from host for degradation for suppressing PTI
AvrPia	85	Cytoplasm	Interaction with *Pi-a*
AvrPii	70	Cytoplasm	Interaction with *Pi-i*
AvrPik/km/kp	113	Cytoplasm	Interaction with *Pi-k*
AvrPib	75	Cytoplasm (predicted)	Interaction with *Pi-b*
Puccinia graminis f. sp tritci (biotroph; hosts: wheat and barley)			

(Table 1) cont.....

Effector Protein	Length (*amino acid residues*)	Localization in the Plant	*R*-protein association; Role in Virulence
RGDBP	818	Apoplast	Interaction with *Rpg1*
VPS9	744	Apoplast	Interaction with *RpgA*
PGTAUSPE-10-1	Unknown	Unknown	Possible interaction with *Sr22*
***Ustilago hordei* (biotroph; host: barley)**			
UhAvr1	170	Not reported	Interaction with *Uh1*
***Ustilago maydis* (biotroph; host: maize)**			
Pep1 (protein essential for penetration1)	178	Apoplast	Required for full virulence; Directly interacts with *POX12* and suppresses plant defense by scavenging ROS
Pit2	118	Apoplast	Required for full virulence; Directly interacts with cysteine protease
Cmu1		Apoplast	Required for full virulence; A chorismate mutase that potentially cooperates with maize chorismate mutase to reduce available SA
Tin2 (Tumor inducing 2)	207	Apoplast	Required for full virulence; Interacts with maize *TKK1* which positively influences anthocyanin production
See1	157	Expressed in nucleus and cytoplasm	Required for full virulence
***Phytophthora infestans* (oomocyte; hosts: potato and tomato)**			
Avr2	118	Cytoplasm	Interaction with *R2*
Avr3a	147	Cytoplasm	Interaction with *R3a*; INF1-meidated plant death
Avr4	287	Likely cytoplasm	Interaction with *R4*
AvrBlb1	153	Cytoplasm	Interaction with *RpiBlb1*; Disrupts cell wall-plasma membrane adhesion
AvrBlb2	101	Cytoplasm	Interaction with *RpiBlb2*; Required for full virulence; blocks plant protease secretion

3. PLANT RESISTANCE (*R*) GENES

The defense of plants by effector-triggered immunity (ETI) involves a very specific interaction between the products of pathogen avirulence (*Avr*) genes and the products resulting from the expression of the host plant resistance (*R*) genes, according to the gene-for-gene model (Flor 1971). *R* proteins can recognize pathogenic effectors directly or indirectly. Indirect recognition occurs in accordance with the so-called guard hypothesis (Dangl and Jones 2001). Currently, two variants of this hypothetical model are recognized. In one, the *R* receptor is constitutively associated with a host intermediate factor (guard protein), while in the other, the effector of pathogens first associates with a target guard protein, and the complex formed is then recognized by the receptor of the plant immune system (Elmore *et al.* 2011). The major proof of the guard hypothesis was obtained in the R/Avr system between *A. thaliana* and *Pseudomonas syringae* pv. *tomato*, where modification of the host factor RIN4 by the product of the bacterial *Avr* gene activates the resistance protein RPM1 in *A. thaliana*, resulting in plant resistance (Mackey *et al.* 2002).

Structurally, *R* gene products generally have a central nucleotide-binding site domain (***Nucleotide Binding Site domain***, *NBS*), a C-terminal region rich in leucines (***Leucine-Rich Repeat***, LRR) and a variable N-terminal domain, mainly identified as TIR (Toll / Interleukin-1) or CC (Coiled-coil) (Gururani *et al.* 2012). Besides TIR-NBS-LRR and CC-NBS-LRR, other major classes of *R* genes include RLK (containing an extracellular LRR, a transmembrane domain and a cytoplasmic kinase domain), RLP (which are similar to RLK but do not have the kinase domain) and cytoplasmic enzymatic *R* proteins, which contain neither LRR nor NBS (Gururani *et al.* 2012) [5].

4. ELEMENTS OF DIFFERENTIATION BETWEEN PTI AND ETI

In general, the signaling pathways leading to PTI or ETI, as well as defense reactions (oxidative burst, activation of MAPK, transcriptional reprogramming, modification of the hormonal balance, secretion of antimicrobial compounds, *etc.*)[6] are the same. HR is often seen as a characteristic reaction to ETI, but it is sometimes seen in the context of PTI. For example, the recognition of flagellin from *P. syringae* pv. *tabaci* induces HR, as does the recognition of the CBEL glycoprotein from *Phytophtora parasitica* (Khatib *et al.* 2004; Naito *et al.* 2008).

However, what differentiates mainly PTI and ETI is the ***kinetics and magnitude of responses***. Indeed, the responses triggered in the event of an ETI are stronger and more prolonged. At least three studies can be indicated in this context:

a. Eulgem (2005) reported that MAPK cascades, but also signaling pathways mediated by plant hormones, experience very significant reprogramming of defense gene expression during ETI and PTI [7]. This corresponds to 25% of the genes in *A. thaliana*, which are significantly deregulated during infection. If the genes deregulated during ETI and PTI were broadly the same, this deregulation proved to be much ***more prolonged*** in the context of ETI.

b. On the other hand, Torres *et al.* (2006) demonstrated that the production of ROS following the recognition of a PAMP is rapid and transient, whereas the production of ROS caused by the recognition of an effector is ***biphasic***: a first peak of transient and low production is observed, then a second peak appears, more durable and of stronger amplitude.

c. Likewise, Tsuda *et al.* (2009) developed a model of PTI and ETI responses in which these two types of immunity differ kinetically: the reactions of PTI are ***synergistic***. Each immune sector activates the others, and the combined effect of all sectors is greater than the sum of their individual effects. Therefore, there is a positive interaction between immune sectors. Rather, the ETI induces ***compensatory responses***. The loss of one sector is immediately compensated by the others. This model explains why ETI is much more difficult to defeat by pathogen effectors. PTI is often surmounted, because an effector only needs to delete a sector, resulting in the rest of the immune network being strongly impacted and the synergistic effect lost. In contrast, in ETI, the loss of one sector is easily compensated for by the other sectors (or subnetworks). These relationships therefore explain the ***vulnerability*** of the resistance induced during PTI *versus* the ***robustness*** of the resistance conferred by ETI. This study of Tsuda *et al.* (2009) was conducted taking into account 3 sectors (SA, JA and ET) [8], but many other sectors are to be considered in plant immunity (MAPK, auxin, gibberellins, abscissic acid ...).

CONCLUSION

Effector-triggered immunity (ETI) was originally termed gene-for-gene resistance and dates back to fundamental observations of flax resistance to rust fungi by Harold Henry Flor in the 1940s. Since then, genetic and biochemical approaches have defined our current understanding of how plant resistance proteins recognize microbial effectors. ETI is a system which is more amplified and faster than PTI and which usually develops to trigger the hypersensitive response (HR) leading the infected host cell to apoptosis. In recent years, significant research advances have revealed the host defense processes, the specific host proteins targeted by pathogens, the mechanisms of *R* protein activation, and how pathogen and *R* protein effectors co-evolve. These advances have practical ramifications for resistance durability and for future resistance engineering.

[1] In the model species *Arabidopsis thaliana*, RRS1 stands for « resistance to *Ralstonia solanacearum* 1, while PopP2 is an *R. solanacearum* effector. *R. solanacearum* is the causal agent of bacterial wilt in a wide range of host species including agronomically important Solanaceae species.

[2] Flax rust (*Melampsora lini*) AvrL567 avirulence proteins and the corresponding flax (*Linum usitatissimum*) L5, L6, and L7 resistance proteins interact directly

[3] Chapter 10 of this book is entirely devoted to the study of the different, direct or indirect, molecular models of host-pathogen recognition.

[4] An extensive inventory of ~83 effector proteins cloned and characterized from crop-infecting fungi and oomycetes is available in the review paper of Selin *et al.* (2016) with bibliographical references of the original studies.

[5] In this context, consult Chapter 13 of this book, dedicated to the molecular classification of disease resistance genes in plants.

[6] Refer to chapters 15, 16 and 18 of this book for further investigation.

[7] See Chapter 16 of this book for more information on transcriptional reprogramming associated with the plant's immune response.

[8] It is to note that these three sectors, which are important for immunity against necrotrophic and biotrophic pathogens, are typically antagonistic.

CHAPTER 8

Acquired Resistance and Elicitors of Natural Plant Defense Mechanisms

Abstract: Unlike innate resistance, acquired resistance is a defense system activated mainly by an earlier infection and allowing plants to resist later attacks by harmful organisms. Its mode of action does not depend on the direct destruction or inhibition of the invading pathogen, but rather on physiological changes which lead to the increase of the physical or chemical barrier of the host plant. The idea of using this ability of plants to defend themselves, to the aim of protecting them from their bio-aggressors is a completely realistic strategy that can be reached by using certain molecules, which have eliciting properties. These molecules, called natural defense stimulators (NDSs), can be of natural or synthetic origin and are capable of putting the plant on a state of alert in order to respond quickly and effectively in subsequent attacks. This innovative strategy greatly contributes to reducing the risks associated with pesticides, and also has great promises for the future, in terms of both socio-economic impact and technology transfer. This chapter provides a summary of the remarkable progress made in recent years in understanding the mechanisms involved in the acquired resistance of plants to various pathogens.

Keywords: Elicitors of natural plant defense, Induced Systemic Resistance (ISR), Jasmonic Acid (JA) et Ethylene (ET) signalling, Local Acquired Resistance (LAR), Pathogenesis-Related (PR) Proteins, Plant growth promoting rhizobacteria (PGPR), Salicilic Acid signalling, stress memory, Systemic Acquired Resistance (SAR).

INTRODUCTION

Plants are sessile organisms that grow within complex, sometimes adverse environments, and are always faced with many stresses of a biotic and abiotic nature. These stresses can lead to significant reductions in crop yield and quality (Atkinson *et al*. 2011; Singh *et al*. 2002). In order to overcome these constraints, plants have developed *innate defensive strategies* that allow them to resist different types of threats. In general, there are two types of defense in plants: *passive defense*, [1] involving preformed or constitutive barriers which the plant has acquired following environmental adaptations (*e.g.* the cuticle), and *active resistance*, involving instantly formed barriers in response to stress (Llorens *et al*. 2017). When a pathogen succeeds in bypassing the first line of passive defense,

Dhia Bouktila & Yosra Habachi

the active resistance system will be set up, leading to considerable modifications in the metabolic activity of plant cells, resulting in a cascade of events intended to restrict the progression of the infectious agent (Benhamou and Rey 2012). First, broad spectrum defenses are induced, which are often linked to the detection by plants of pathogen-associated molecular patterns (PAMPs) (Jones and Dangl 2006). This is called ***PAMP-triggered immunity (PTI)***[2]. Several membrane proteins involved in the recognition of PAMPs have been identified, in particular the FLS2 protein in *Arabidopsis*, a receptor associated with the perception of flagellin-type bacterial proteins (Chinchilla *et al.* 2006). Plants are also capable of responding to the presence of pathogens *via* the detection of pathogen effectors; in this case, they are using ***effector-triggered immunity (ETI)*** (Jones and Dangl 2006) [3].

Apart from innate immunity (basal and race-specific), ***acquired resistance*** occurs when the inoculation of a plant with an incompatible pathogen strain (one that does not succeed to invade the host plant because of either race-specific or basal resistance) induces a plant defense response that prevents infection by a subsequent inoculation with a normally virulent pathogen.

1. ACQUIRED RESISTANCE

The defensive strategies adopted by plants, in the event of a pathogenic attack, generally result in a hypersensitivity reaction (HR), which is, in fact, a programmed death of plant cells in the area of infection (Jones and Dangl 2006). Following this reaction (HR), signals are sent to uninfected cells and this contributes to the establishment of a ***permanent standby state*** that allows the plant to be alert in the event of a potential attack and to respond quickly to aggression (Jourdan *et al.* 2008); it is, therefore, an ***acquired resistance***. This resistance takes place, first of all, over a slightly wider area than that infected, this is the case of ***local acquired resistance (LAR)***, which is characterized by an increase in the resistance of tissues adjacent to the infection site (Kombrink and Schmelzer 2001). Furthermore, the resistances implemented by plants are not limited to local responses only; plants are also capable of deploying systemic resistance, that is to say generalized to all of their tissues (Durrant and Dong 2004; Jourdan *et al.* 2008). This is the case with S*ystemic Acquired Resistance (SAR)* and *Induced Systemic Resistance (ISR)*.

1.1. Systemic Acquired Resistance (SAR)

Acquired Systemic Resistance (SAR), also known as ***Acquired Physiological Immunity***, concerns the plant as a whole (Durrant and Dong 2004). Since the

1930s, Chester has already mentioned the possibility that acquired physiological immunity may exist in plants (Chester 1933). However, it was only after the 1960s that Ross took up this concept. He found that following a first infection of tobacco plants (*Nicotiana tabacum* L.) by the tobacco mosaic virus (TMV), a second infection caused less damage to the whole plant (Ross 1961a; 1961b). Generally, SAR appears throughout the plant from 30 minutes to several hours after initial infection and results in a much more effective defense in subsequent attacks. Even more remarkable, this exacerbation of defenses is effective against any pathogen (viruses, bacteria or fungi), and not just that involved in the first attack (Klarzynski and Fritig 2001). This resistance can last for several weeks. In cucumbers, for example, it has been shown that after the first inoculation with a pathogen, the plants were protected until flowering (Madamanchi and Kuc 1991). Generally, this type of resistance is accompanied by the accumulation of *salicylic acid (SA)* and *Pathogenesis-Related Proteins (PR proteins)* (Fig. 1). Indeed, the accumulation of salicylic acid after infection is necessary to activate pathogenesis-related (PR) defense genes (Vallad and Goodman 2004). For example, assays of salicylic acid have shown that its concentration increases considerably (x 100) in several plants following a pathogenic infection (Yalpani *et al.* 1993). In addition, it has been reported that transgenic tobacco and *Arabidopsis thaliana* L. plants unable to synthesize salicylic acid no longer express SAR against various pathogens (viral, fungal and bacterial) (Delaney *et al.* 1994 ; Dong 1998). However, recent studies have highlighted several metabolites that may be involved in *long-distance SAR signaling*, such as dehydroabietinal (DA), azelaic acid (AzA) and pipecolic acid (Pip) (Dempsey and Klessig 2012; Shah and Zeier 2013).

Disease Acquired Resistance is a paradigm for the existence of a form of *plant memory*. This state of resistance can be compared to vaccination practiced in humans, with the difference that it protects against many pathogens of different natures.

1.2. Induced Systemic Resistance (ISR)

The protective effect conferred by SAR is phenotypically similar to another phenomenon triggered by interaction with a *non-pathogenic microorganism*. This immunization of the plant is called induced systemic resistance (ISR) (Bakker *et al.* 2013); in fact, it is a form of induced resistance specifically stimulated by rhizobacteria better known as *Plant growth-promoting rhizobacteria* (PGPR). These include Gram$^+$ bacteria such as *Bacillus pumilus*, *B. subtilis*, and *B. thuringiensis* (Kloepper *et al.* 2004), or Gram$^-$ bacteria, most belonging to the genus *Pseudomonas*, which are most studied in the context of

ISR (Jourdan *et al.* 2008). During the 1980s, PGPRs attracted most attention because of their ability to stimulate plant growth (Kloepper and Schroth 1981). Then, the role of these PGPRs in the induction of a systemic resistance against pathogenic agents, was demonstrated for the first time by Scheffer (1983) who noted that a pretreatment of the American elm (*Ulmus americana* L.) with strains of *P. fluorescens* resulted in a significant reduction in symptoms caused by the fungus *Ophiostoma ulmi*. Since then, PGPRs have also been considered as biological control agents. Subsequently, several descriptive studies have also shown that certain strains of these bacteria (in particular *Pseudomonas* spp.) have the capacity to reduce the impact of several root diseases in a multitude of cultivated plants, both mono- and dicotyledonous (Kloepper 1991 ; Paulitz *et al.* 1992). For example, the application of *Bacillus* spp. on rice protects the plant from leaf blight (Chithrashree *et al.* 2011). Generally, the ISR resistance mechanisms reach their maximum effectiveness between 4 to 5 days after the application of the inducing agent, but the level of resistance persistence decreases over time. In addition, the durability of resistance induced by PGPRs differs from one plant to another, and according to the bacterial strain used (Ramamoorthy *et al.* 2001). In contrast, ISR differs from SAR in that it is ***independent of the SA pathway and does not activate the PR genes***, but depends on the signaling pathways of ***jasmonic acid (JA)*** and ***ethylene (ET)*** (Fig. **1**). This was confirmed when *A. thaliana* mutants for JA and ET response did not show increased resistance against *P. syringae* after colonization by PGPR bacteria (Vleesschauwer and Hofte 2009).

1.3. Metabolic Changes Associated with Induced Resistance

Most studies have shown great similarities between very diverse plant species, during induced resistance, with three major categories of metabolic changes systematically found and which are costly in energy terms (Klarzynski and Fritig 2001):

a. ***Production of large quantities of plant antibiotics, especially phytoalexins*** (Ongena *et al.* 2000), which are considered to be key components of induced resistance in plants. These compounds accumulate in the infected tissues of the plant, thereby reducing or stopping the progression of the pathogen by disorganizing and killing its cells (Jourdan *et al.* 2008).
b. ***Strengthening of the cell wall*** (Soylu 2006), by depositing newly formed macromolecules such as proteins, glycoproteins, polysaccharides, aromatic polymers of lignin type (Klarzynski and Fritig 2001). These deposits will therefore give the extracellular matrix a very high resistance to enzymes of pathogenic agents, which can limit the spread of the infection.

c. ***Production of a wide range of defense proteins.*** Among these are pathogenesis-related (PR) proteins. These proteins are encoded by the host plant and are specifically induced when the plant is subjected to pathogenic infections or any undesirable stress (insects, injuries, exposure to aggressive chemicals or atmospheric conditions). They not only accumulate locally in infected cells but are also systemically induced. These PR proteins disarm the attacker by acting as antifungal and / or antibacterial response elements that are capable of damaging the walls of fungal or bacterial cells (Andreu *et al.* 2006; Chandrashekar and Satyanarayana 2006).

All these molecules play a crucial role in the plant's defensive response and act synergistically in a context where the speed and amplitude of the overall response are key parameters in the future of the plant-pathogen relationship.

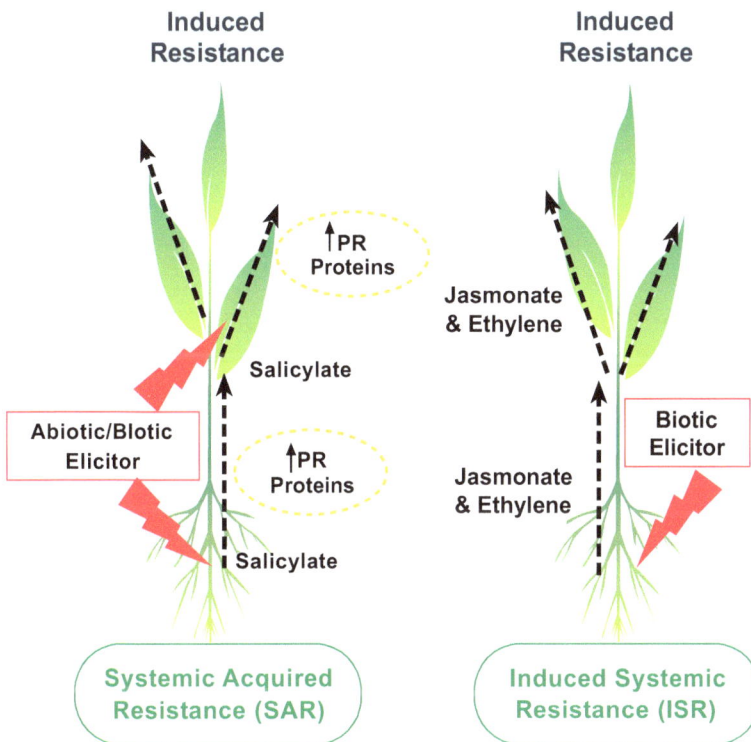

Fig. (1). Schematic view of systemically-induced immune responses. Acquired systemic resistance (SAR), induced by exposure of root or leaf tissues to abiotic or biotic elicitors, depends on salicylic acid and is associated with the accumulation of pathogenesis–related (PR) proteins. Induced systemic resistance (ISR), induced by exposure of the roots to specific strains of rhizobacteria that promote plant growth, depends on ethylene and jasmonic acid, independently of salicylate, and is not associated with PR proteins accumulation (modified from Vallad and Goodman 2004).

2. ELICITORS OF NATURAL PLANT DEFENSE MECHANISMS. CAN PLANTS BE IMMUNIZED?

A number of recent studies have revealed that multiple natural products are capable of playing a role of elicitors to generate a delayed response to pathogens, and the different chemical nature of these elicitors suggests that there is no single structural feature determining the eliciting mechanism (Cragg and Newman 2013). These spectacular advances in the knowledge of the mechanisms involved in induced resistance in plants have been a starting point for initiatives aimed at finding new elicitors to protect crops. This implies *translating the natural defense strategies of plants into application* in a context that combines environmental preservation (reduction of risks linked to pesticides) and intensive production of agricultural products. This essentially preventive plant protection strategy consists of external treatments with substances capable of stimulating natural defenses. These substances are generally called *natural defense stimulators (NDS)* (Benhamou 2009). Today, an increasing number of natural defense stimulators (NDSs) are coming onto the market, notably Stifénia®, Milsana®, Iodus 40®, Elexa™ and Messenger®, and it is reasonable to believe that more and more *resistance inducers* will be available in the near future (Benhamou and Rey 2012).

The preventive application of NDS on a plant aims to enable it to defend itself against its most harmful future enemies (pathogens and pests). The objective of this strategy is therefore to put the plant on a state of alert in order to respond more quickly and more effectively, during subsequent attacks by any pathogenic agent (Benhamou 2009). In terms of molecular response, NDSs induce a large number of physical and biochemical changes, such as the synthesis of defense-associated enzymes and the strengthening of the cell wall, leading to the expression of resistance in a plant (Fig. **2**). It is therefore a kind of *vaccine* capable of activating the immune system of the plant so that a plant initially sensitive to a pathogen becomes resistant.

2.1. Definition of an NDS

By definition, an NDS or elicitor, is a (natural or not) substance, capable of inducing (or preparing for induction), in treated plants, a state of resistance to biotic and/or abiotic stresses (Benhamou and Rey 2012). In NDSs of natural origin such as Stifénia®, Milsana®, Iodus 40®, Elexa ™ and Messenger®, active ingredients are simple molecules that are isolated from certain plants or algae, or even extracted from crustaceans (*e.g.* chitosan[4]) or microorganisms (*e.g.* harpins[5], cerebrosides [6]) (Aziz *et al.* 2003; Barka *et al.* 2004).

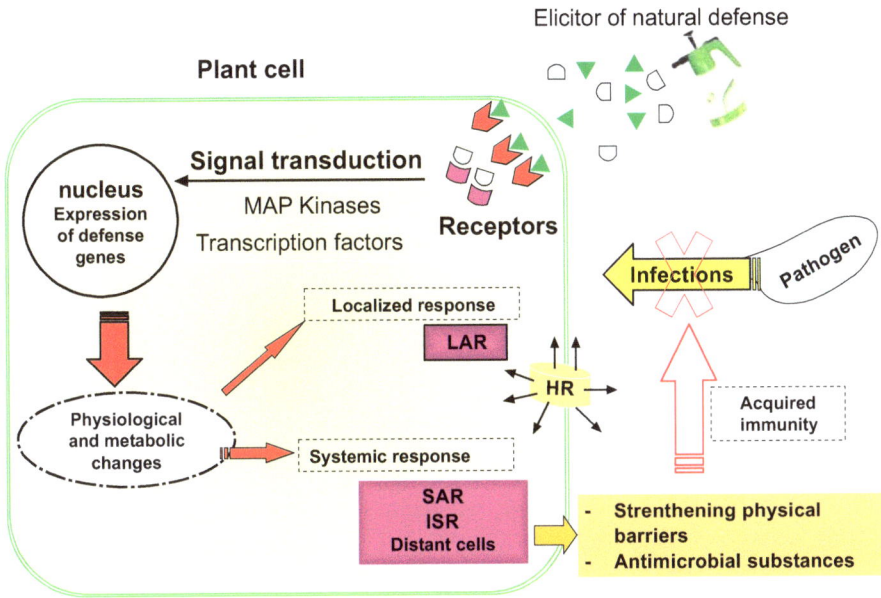

Fig. (2). Mode of action of NDSs.
The application of NDSs allows the plant to acquire systemic resistance, which is most often manifested by the synthesis of a wide range of defense proteins and the strengthening of cell walls, *via* the activation of different metabolic pathways (salicylic acid SA, jasmonic acid JA and ethylene ET, *etc.*). This strategy will allow the plant to defend itself quickly and effectively in subsequent attacks.

For synthetic NDSs, *acibenzolar-S-methyl (ASM)* is the first product developed, in the 1980s. It is a structural analog of salicylic acid. Acibenzolar-S-methyl (ASM) enhances the production of pathogenesis-related proteins (PR)-proteins and causes systemic acquired resistance in treated plants, since it mimics the role of salicylic acid. It was marketed by Syngenta in the United States of America under the trade name Bion® (Benhamou and Picard 1999; Benhamou and Rey 2012). This product has the advantage of being less toxic in external application and of being better transported and distributed in plants than salicylic acid itself (Klarzynski and Fritig 2001). Today it is used on wheat (*Triticum aestivum* L.), lettuce (*Lactuca sativa* L.), spinach (*Spinacia oleracea* L.), radish (*Raphanus sativus* L.) or tomato crops (Benhamou and Rey 2012).

In addition to acibenzolar-S-methyl (ASM) (an SA analog), several other chemical inducers have been synthesized and used as natural defense stimulants in plants to protect them against a wide range of diseases. A nonexhaustive list (reviewed in Zhou and Wang 2018) would include:

• benzothiadiazole S-methyl ester (BTH) (SA analog)
• 2.6-dichloroisonicotinic acid (INA) (SA analog).

- acetylsalicylic acid (aspirin) and methyl SA (MeSA) (SA derivatives)
- nitric oxide (NO),
- dicarboxylic acid azelaic acid (AzA)
- the phosphorylated sugar glycerol-3-phosphate (G3P)
- the abietane diterpenoid dehydroabietinal (DA)
- the amino-acid derivative pipeolic acid (Pip),
- and N-hydroxypipecolic acid (NHP)

2.2. Advantages and Disadvantages of Using NDSs

The development and use of NDSs has several advantages (Lyon *et al.* 1995; Benhamou 2009). These can be synthesized as follows:

a. NDSs are capable of conferring on plants a systemic and lasting (durable) resistance.
b. NDSs have a very broad potential for action. Indeed, they act on a large number of plant species and the defense responses induced in these plants are effective on an equally wide range of pathogens (fungi, viruses, bacteria, *etc.*) (Benhamou and Rey 2012).
c. Their toxicity to the environment is reduced or even non-existent.
d. NDSs are biodegradable.
e. The application of NDSs is relatively simple and integrates perfectly with usual cultural practices.
f. The accumulation of phytoalexins in elite cultivars (treated by NDS), would be rather beneficial for the consumer. Indeed, fruits and vegetables from cultures treated with NDSs were found slightly richer in phenolic compounds (Treutter 2000), in particular flavonoids which have recognized therapeutic properties such as anticancer, antimicrobial, anti-inflammatory and / or cardio-protective (Eberhardt *et al.* 2000; King and Young 1999).

The fact remains that NDSs have many limits to take into account. Among the main problems related to the use of NDSs (Benhamou and Rey 2012):

a. Variability in terms of performance.
b. The energy cost linked to the induction of resistance could result in a slight drop in productivity (yield).
c. An ***allergenic power*** of certain defense proteins, in particular PR proteins (Ebner *et al.* 2001; Gao 2019). Stability and resistance to proteases make these PR proteins excellent candidates for triggering an allergic response (Sinha *et al.* 2014). It is, therefore, possible that plants treated with NDSs (and which

show increased expression of PR proteins) are also associated with increased *allergenicity*, which raises a serious question as to their commercial acceptability.

CONCLUSION

In recent years, the rapid development of agriculture has favored the emergence of certain diseases leading to significant economic losses. Generally, disease control and prevention strategies have been based on the use of chemicals. Today, as part of the risk reduction programs linked to these products, the development of innovative, reliable and environment-friendly strategies is generating growing interest among horticultural and agricultural producers. In this context, the remarkable progress made, in recent years, in terms of understanding the mechanisms involved in induced resistance in plants, has enabled the development and marketing of an increasingly large number of biological molecules capable of stimulating the Immune system of plants by mimicking the effects of pathogens. These molecules are called natural defense stimulators or elicitors, and are capable of triggering a series of biochemical events leading to the expression of resistance in the plant. Despite the obvious advantages of NDSs, further research is still needed to elucidate certain obscure areas, mainly concerning the time required for resistance to be fully expressed, the optimal dose of active ingredient required and, above all, the potential impact on human health.

[1] Refer to Chapter 5 for further documentation.

[2] Refer to Chapter 6 for further documentation.

[3] Refer to Chapter 7 for further documentation.

[4] Chitosan is a marine polysaccharide that forms structural components in the exoskeleton of crustaceans (Dima *et al.* 2017)

[5] Harpins are multifunctional proteins secreted by Gram-negative plant-pathogenic bacteria (Choi *et al.* 2013).

[6] Cerebrosides (ceramide monohexosides, CMHs) can be found in plants, fungi, and animals, with some structural differences between these organisms; yet, cerebrosides are more abundant in fungi, as they seem to be present in almost all fungal species studied so far (Barreto-Bergter *et al.* 2011).

CHAPTER 9

Quantitative Resistance

Abstract: If breeders use a limited number of genes in their new resistant varieties, the adaptive capacity of pathogenic populations will ultimately lead to more or less rapid overcoming of these resistances, thus limiting the sustainability of their effectiveness. Qualitative resistance is considered less durable than quantitative resistance since the latter oppose less selection pressure on the pathogen and they are often governed by several resistance genes. Quantitative disease resistance has been observed within many crop plants but is not as well understood as qualitative (monogenic) disease resistance and has not been used as extensively in breeding. Mapping quantitative trait loci (QTLs) is a powerful tool for genetic dissection of quantitative disease resistance.

Keywords: Additive effect, Marker-Assisted Selection, Molecular markers associated to QTLs, Quantitative trait loci (QTL), Resistance durability.

INTRODUCTION

Plant pathogens are major limiting factors in crop production and this has led to the extensive use of chemicals to control them. Plant genetic resistance is a promising key alternative to control crop diseases and pests. However, pathogens frequently adapt to and overcome genetic resistance especially when it is determined by major genes. The mechanism for bypassing qualitative resistance is explained by the existence of gene-for-gene interactions between the plant's resistance gene and the pathogen's avirulence gene, since a simple mutation in the gene for avirulence can allow it to escape recognition by the resistance gene and overcome this resistance. Thus, quantitative resistance has gained interest in recent years to address the major challenge of genetic resistance durability. Several genes usually control quantitative resistance and are associated with genomic regions or QTL (quantitative trait loci) which contribute, each with variable effect, to the phenotype of resistance to a pathogen. However, a combination of resistance QTLs can lead to total resistance in some cases, especially when QTLs have strong effects (Niks *et al.* 2015). For example, three QTLs, *rx1, rx2*, and *rx3* were found to confer a high level of resistance of tomato to *Xanthomonas campestris* (Stall *et al.* 2009). Over the past 20 years, since the development of molecular markers, many resistance QTL detection experiments have been conducted in all major crop species (Stall *et al*. 2009; Huang and Han 2014; Desgroux *et al*. 2016; Corwin and Kliebenstein 2017).

1. MOLECULAR MECHANISMS ASSOCIATED WITH QUANTITATIVE IMMUNITY

While the molecular mechanisms underlying the main *R* genes have been widely described (Michelmore *et al*. 2013), those associated with quantitative immunity are much less known (Fig. **1**).

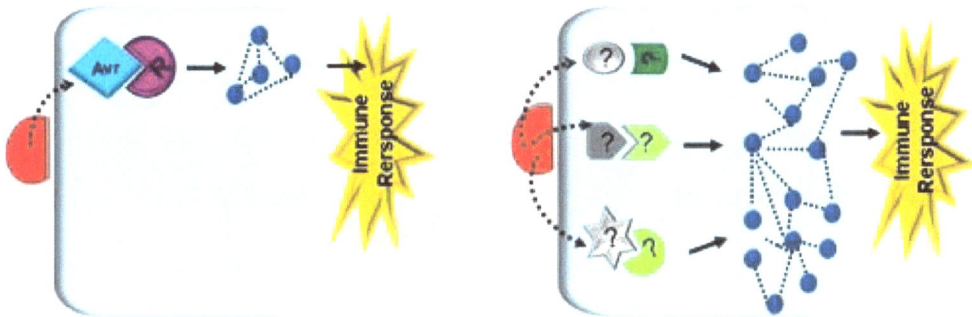

Fig. (1). Simplified model of the molecular bases of plant-pathogen interaction in the case of qualitative resistance (left) and the complexity of molecular bases assumed in quantitative resistance (right) (Red = pathogen cell, blue = plant cell).

The genes that underlie resistance QTLs are far less known and described than the *R* genes. Few resistance QTLs have been cloned to date and have shown a variety of underlying functions (Niks *et al*. 2015). The main QTLs cloned by positional cloning are *Lr34, Yr36, Pi21, Rhg1, Rhg4* and *Pi35* (Lavaud 2015). Their recognized functions are disparate, which include ABC transporter proteins, tRNA protein (HIS) guanylyltransferase, Proline-rich protein, Serine hydroxymethyltransferase or NBS-LRR domain-containing proteins (Kou and Wang 2010 ; Michelmore *et al*. 2013; Niks *et al*. 2015). These QTL cloning results fuel the multiple *hypotheses* synthesized by (Poland *et al*. 2009) on the *function of resistance QTLs*. Among these:

a. Resistance QTLs would be correlated to the plant morphology and development.

    Many correlations have been described between QTLs of resistance, on one hand, and architecture and development of the plant, on the other hand. In potato, colocalizations have been observed between QTLs of resistance to

Phytophthora infestans and plant maturity and vigor (Collins *et al.* 1999; Gebhardt *et al.* 2004). In pea, QTLs of partial resistance to *Didymella pinodes[1]* have been located in the same regions as genes controlling the elongation of internodes (plant height), flowering date and photoperiod sensitivity during the initiation of flowering (Prioul *et al.* 2004; Giorgetti 2013).

b. Resistance QTLs would be involved in the production of antitoxic compounds.

Biochemical studies of the *Arabidopsis-Botrytis* pathosystem have shown that camalexin[2] levels are correlated with the level of quantitative resistance. In fact, the decrease in the level of camalexin in the plant caused greater sensitivity towards the pathogen (Denby *et al.* 2004).

c. Resistance QTLs would be involved in the transduction of the plant defense signal.

In *A. thaliana*, mutants of the transcription factor WRKY (involved in the signaling pathways regulated by salicylic acid and jasmonic acid) showed an increased sensitivity to the necrotrophic pathogen *Botrytis cinerea* (Zheng *et al.* 2006). Conversely, increased resistance to *Pseudomonas syringe* pv. tomato, and *Peronospora parasitica* has been observed in mutants of *A. thaliana* in MAP kinase 4 protein (MPK4) (Petersen *et al.* 2000). In *A. thaliana*, QTL RKS1 confers resistance to *Xanthomonas campestris* and corresponds to a protein kinase potentially involved in signal transduction (Huard-Chauveau *et al.* 2013).

d. QTLs would be *R* genes with weak effects or *R* genes overcome by pathogen strains yet maintaining a residual effect.

Several studies have shown colocalization of QTLs and overcome *R* genes in several plants (Poland *et al.* 2009). In corn, QTL RCG1 has been identified as an NB-LRR gene involved in partial resistance to anthracnose (Broglie *et al.* 2006). In rice, substitutions in the LRR domain of the *Xa21* gene have conferred partial resistance to *Xanthomonas oryzae* pv. *oryzae* (Wang *et al.* 1998). In wheat, the *Yr36* gene conferring quantitative resistance to yellow rust is involved in the transfer of lipids and includes a lipid-binding domain (Fu *et al* 2009). The partial effect of a QTL could therefore come from the modification of recognition sites and / or transducing sites, leading to a weak interaction between the pathogen and the plant and consequently, would trigger less important and/or slower defense responses.

e. QTL would be unique genes never identified before.

This hypothesis is supported by recent QTL cloning results, which have identified new classes of genes. In rice, a study on resistance to *Magnaporthe grisea* identified a region of 65 kb, underlying the QTL *Pi34* containing 10 open reading frames (ORFs) whose gene function is unknown (Zenbayashi-Sawata *et al.* 2007).

2. BREEDING FOR QUANTITATIVE RESISTANCE

Because of their efficiency and durability, ***resistance QTLs*** are good candidates for future breeding programs, both for their spectrum of action and for their sustainability. However, the polygenic nature of partial resistance makes the management of breeding programs more complicated, compared to specific resistance breeding. The selection of this type of resistance (quantitative) is a long and expensive process compared to the selection of monogenic resistance genes. QTL mapping requires populations of large descendants, with enough markers to saturate genetic maps associated with fine and reliable phenotyping.

Generally, it is difficult to predict the usefulness of QTL for marker-assisted selection based only on the QTL performance in a single genetic background and limited number of environments, tested in a particular study. Therefore, (Goffinet and Gerber 2000) proposed a method, called ***QTL meta-analysis***, which combines QTL results from different independent analyses. These QTL regions from the QTL meta-analysis ('meta-QTL', or MQTL) provide refined (or consensus) positions, and they can facilitate the identification of positional candidate genes.

In recent years, meta-analysis of QTL/resistance genes to multiple bio-aggressors has been carried out in several species, showing that resistance QTLs can be distributed throughout the genome. In rice, QTL meta-analysis made it possible to identify colocalizations of QTLs and *R* genes for resistance to *Magnaporthe oryzae* as well as a choice of QTL interesting for future breeding programs (Ballini *et al.* 2008). In *Solanaceae*, several studies of comparative QTL of resistance in potato, tomato and pepper have demonstrated the conservation of the position of major resistance QTLs (Grube *et al.* 2000; Pan *et al.* 2000).

3. SPECIFICITY OF QTLS

The effects of quantitative resistance are often ***additive***. However, QTLs with non-additive, for example ***epistatic***, effects have been identified (Lefebvre and Palloix 1996; Manzanares-Dauleux *et al.* 2000; Calenge *et al.* 2005). Quantitative resistance can sometimes only be detected under certain environmental conditions (soil, climate, population of pathogens), or in specific genetic contexts. Thus, stable QTLs are highly sought after for their applicability in transgenesis (Calenge and Durel 2006; Ballini *et al.* 2008; Danan *et al.* 2011; Hamon *et al.* 2011; Goudemand *et al.* 2013).

QTLs that are detected in a single plant genotype are called genotype-specific QTLs, by oppsosition to non-specific (or broad spectrum) QTLs. In several systems, quantitative resistance has been found as a result of the combination of broad spectrum QTLs and QTLs specific to one or more isolates (Caranta *et al.* 1997; Calenge *et al.* 2004; Rocherieux *et al.* 2004).

A QTL can also be specific or non-specific to a species of pathogen, and some QTLs can confer resistance to multiple pathogens (Ellis *et al.* 2014; Wiesner-Hanks and Nelson 2016).

4. RELATIONSHIP BETWEEN GENES, PROTEINS, METABOLITES AND QTL

Genes, proteins and metabolites regulated by quantitative resistance in response to infection by pathogens have also been identified using the transcriptomic and metabolomic approaches (Kushalappa *et al.* 2016). For example, metabolites involved in antimicrobial degradation and cell wall strengthening activities have been identified in several plant species. Indeed, in potato, quantitative resistance to late blight was mainly associated with the thickening of the cell wall due to the deposition of hydroxycinnamic acid, flavonoids and alkaloids (Yogendra *et al.* 2015). In *A. thaliana*, quantitative resistance to clubroot infection[3] is considered to result mainly from the segregation of multiple alleles (Liégard 2018). A major phytoalexin, camalexin[4], induced in response to clubroot infection, is associated with reduced development of pathogens in lines carrying a major resistance QTL (Lemarie *et al.* 2015).

5. MOLECULAR MARKERS ASSOCIATED WITH QTLS

The development of molecular markers increased researches on assessment, validation and characterization of disease resistant QTLs. However, low-effect QTLs are generally unprofitable in plant breeding, because resistance will have physiological costs for the plant (Brown and Rant 2013). Nevertheless, the development of genomics over the past 10 years has opened up the prospects for *genomic selection of quantitative resistance*, which should lead to better consideration of minor-effect QTLs in resistance selection programs (Poland and Rutkoski 2016).

In recent years, the development of high-throughput sequencing technologies (NGS) has accelerated the discovery of a larger number of QTLs that provide quantitative resistance. Markers closely linked to QTLs have also been identified in several economically-important crop species (Yu and Buckler 2006; Huang and Han 2014; Desgroux *et al.* 2016). In a recent review, Bartoli and Roux (2017) reported thirty-five studies on plant resistance to genome-wide pathogens (GWAS). In these studies, molecular markers (Single nucleotide polymorphisms, SNPs) closely linked to QTLs have been reported and validated in at least 33 phytopathogenic systems (pathosystems) (Table **1**).

Table 1. Overview of genome-wide association studies in 33 plant pathosystems reviewed in (Bartoli and Roux 2017). [5]

Biotic stress	Pathosystem	Genes Underlying QTLs that have been Functionally Validated
Bacteria	Arabidopsis thaliana / Pseudomonas syringae	RPM1 ;
		RPS5 ;
		RPS2 ;
		AtABCG36
	Arabidopsis thaliana / Pseudomonas viridiflava	No
	Arabidopsis thaliana / Ralstonia solanacearum	RRS1/RPS4 & SSL4
	Arabidopsis thaliana / Xanthmonas campestris	RKS1
		AT5G22540
		RRS1/RPS4
	Glycine max / Xanthomonas axonopodis	No

(Table 1) cont.....

Biotic stress	Pathosystem	Genes Underlying QTLs that have been Functionally Validated
Fungi	Arabidopsis thaliana / Botrytis cinerea	Yes
		No
	Brassica napus / Leptosphaeria maculans	No
	Glycine max / Fusarium virguliforme	No
	Glycine max / Sclerotina sclerotorium	No
	Glycine max / Cadophora gregata	No
	Glycine max / Diaporthe phaseolorum	No
	Glycine max / Phakopsora pachyrhizi	No
	Hordeum vulgare / Puccina graminis	No
	Medicago sativa / Verticillium alfalfae	No
	Oryza sativa / Pyricularia oryzae	No
	Oryza sativa / Magnaporthe oryzae	Pi5 locus
	Setaria italica / Magnaporthe grisea	No
	Setaria italica / Rhizoctonia solani	No
	Setaria italica / Uromyces setariae-italicae	No
	Sorghum bicolor / Macrophomina phaesolina	No
	Sorghum bicolor / Fusarium thapsinum	No
	Zea mays / Cochliobus heterostrophus	No
	Zea mays / Setosphaeria turcica	No
	Zea mays / Cercospora zeae-maydis + C. zeina	No
	Zea mays / Setosphaeria turcica	No
	Zea mays / Sphacelotheca reiliana	No
	Zea mays / Fusarium verticilliodes	No
	Zea mays / Puccina sorghi	No
Oomycete	Arabidopsis thaliana / Hyaloperonospora arabidopsidis	RPP13,RPP5 ; RPP7
	Glycine max / Phytophthora sojae	No
	Medicago truncatula / Aphanomyces euteiches	No
	Pisum sativum / Aphanomyces euteiches	No
	Setaria italica / Sclerospora graminicola	No

6. DURABILITY OF QUANTITATIVE RESISTANCES

It is generally accepted that polygenic resistances are more durable than single major gene resistances (Parlevliet 2002; Palloix *et al.* 2009; Quenouille *et al.*

2013a). Indeed, for a polygenic resistance to be overcome, the pathogen will have to acquire a larger number of mutations than when it is subjected to a selection pressure caused by a single major resistance gene and the probability that all these different mutations will fix permanently in the same genome of the pathogen remains unlikely (Palloix *et al.* 2009). Although quantitative resistance is generally more durable than qualitative resistance, cases of major genes conferring lasting resistance are known. For example, the *Mlo* gene has been used since 1979 to control powdery mildew in varieties of spring barley in Europe (Acevedo-Garcia *et al.* 2014). In addition, several genes conferring resistance to viruses have also shown durability; examples include *Pvr22*, *Pvr4* in pepper (*Capsicum annuum*) [6], and the *Ry* gene in potato (García-Arenal and McDonald 2003). However, in the general case, virulent isolates of the pathogen[7] more or less quickly overcome the main *R* genes (Brown 2015).

The increased durability of quantitative resistance could be due to a number of factors:

a. the effect of partial resistance exerting a *low selection pressure* on the population of pathogens.
b. a combination of contradictory selection pressures on the evolution of pathogens.
c. a low probability of multiple mutations necessary to overcome multiple QTLs.
d. a combination of different mechanisms associated with resistance, which together are more difficult to overcome.
e. and a combination of resistance mechanisms acting successively at different times in the pathogen's life cycle and at different times of the development of plants (Palloix *et al.* 2009; Mundt 2014).

Nevertheless, experimental evolutionary studies have suggested that fungi and viruses are capable of *eroding* and overcoming this type of resistance (Kolmer and Leonard 1986; Lehman and Shaner 1997; Montarry *et al.* 2012). Indeed, the breakdown or erosion of quantitative resistance (or resistance of QTLs) by isolates of pathogens has been observed. Studies of these systems have made it possible to better understand the adaptation processes of the pathogens involved (Cowger and Mundt 2002; Peressotti *et al.* 2010; Caffier *et al.* 2016; Delmas *et al.* 2016).

CONCLUSION

Quantitative resistance appears to be far more durable than qualitative resistance. Even so, an adequate choice of resistance QTLs is still required to create optimal combinations and reduce QTL erosion. The selection of QTLs could be based on their underlying molecular modalities, their spectrum of activity on various pathogen strains, and their effect on pathogen characteristics. In addition, the development of **genome editing** methodologies [8] will open up new perspectives for the creation of novel specificities in resistance genes, which can then be combined to maintain resistance effectiveness.

[1]*Didymella pinodes* (syn. *Mycosphaerella pinodes*) is a hemibiotrophic fungal plant pathogen and the causal agent of ascochyta blight on pea plants.

[2] Camalexin (3-thiazol-2-yl-indole) is a simple indole alkaloid found in *A. thaliana* and other crucifers. This secondary metabolite functions as a phytoalexin to deteriorate bacterial and fungal pathogens.

[3] Clubroot disease, caused by the protist obligate parasite *Plasmodiophora brassicae*, is a major root disease of cultivated Brassicaceae. *P. brassicae* is a soil-borne pathogen that infects the roots of host plants, inducing the formation of galls and negatively affecting plant growth and productivity. Resistance to clubroot in Brassicaceae involves qualitative and quantitative trait loci, as well as epigenetic variation (Liégard et al. 2019).

[4] Camalexin (3-thiazol-2-yl-indole) is a simple indole alkaloid found in the plant *Arabidopsis thaliana* and other crucifers. This secondary metabolite functions as a phytoalexin active against bacterial and fungal pathogens.

[5] The original bibliographical references of each pathosystem are reviewed in Bartoli and Roux (2017).

[6] However, in the pepper-PVY pathosystem, we cannot exclude – as it was revealed by some studies (Quenouille et al. 2013b; 2014) - that the durable Pvr resistance is mostly a result of an additional layer of quantitative resistance, acting as a genetic background with a protective effect for the major Pvr resistance genes.

[7] Also named resistance-breaking (RB) pathogen genotypes.

[8] Documented in Chapter 20.

CHAPTER 10

Molecular Models of Specific Host-Pathogen Recognition

Abstract: Although some of the resistance strategies rely on simple physical or chemical barriers, modern concepts of plant immunity emphasize the role and evolution of protein receptors in the plant cell. These immune receptors, made up of multidomain proteins, are the key elements in the recognition of pathogen elicitors / effectors, leading to the susceptibility or resistance of plants. Numerous pairs of plant R proteins and corresponding pathogenic Avr proteins have been identified as well as cellular proteins which mediate R/Avr interactions, and the molecular analysis of these interactions has led to the formulation of models on how *R* gene products recognize pathogens. Data from several R/Avr systems indicate that specific domains within R proteins determine recognition specificity. However, recent evidence suggests that R proteins have recruited cell recognition cofactors that mediate interactions between Avr proteins and R proteins. Overall, to explain this direct or indirect interaction, at least four models are currently widely approved. This chapter highlights the current trends in understanding host–pathogen interactions through a variety of models.

Keywords: Avirulence (Avr) protein, Compatible/incompatible reaction, Decoy model, Gene-for-gene model, Guard model, Helper NLRs, Integrated decoy model, NLR-IDs, R-Avr recognition, Recognition cofactors, Resistance (R) protein, Resistance gene, Receptor-ligand model.

INTRODUCTION

In their struggle against attacks by viruses, bacteria, fungi, protozoa, nematodes and insects, plants have developed an integrated defense mechanism against the invasion of pathogenic organisms and pests. This integrated mechanism limiting the proliferation of harmful agents to the plant, is called *resistance*. Genes that confer a resistant phenotype are called *resistance genes*. Even though *Flor's model* has been experimentally confirmed in several models, such as *Pita / AvrPita* (Jia *et al.* 2000), *PopP2 / RRS-1* (Deslandes *et al.* 2002), and *L / AvrL567* (Dodds *et al.* 2006), cases of direct *R-Avr* recognition remain rare, suggesting a more complex relationship (Ellis *et al.* 2000).

Since 1998 (Van Der Biezen and Jones 1998b), there has been a new hypothesis called *guard hypothesis*, which involves a third component, a *guardee* protein,

Dhia Bouktila & Yosra Habachi

mediating the interaction during the recognition process. According to this hypothesis, the attachment of the elicitor (Avr) would induce the formation of the guardee-*R* protein complex, causing signal transduction and the defensive response. This hypothesis then explains the fact that certain resistance proteins are capable of recognizing multiple unrelated Avr proteins.

In 2008, Van der Hoorn and Kamoun proposed the ***Decoy model*** in addition to the guard model. They postulated that in most cases, the guardee would not be the target of the effector. A ***decoy protein***, by mimicry with the real target, would trap the effector by competing with its virulence target.

1. « RECEPTOR – LIGAND » MODEL

The first model, called the ***gene-for-gene or receptor - ligand model*** (Fig. **1**), involves the direct effect of a plant receptor that recognizes an effector specific to the pathogen. This model was formalized by Flor (1971), based on studies carried out on rust resistance in flax (*Linum usitatisimum* L.). The initial hypothesis put forward by Flor postulates that for each resistance gene in the host, there is an avirulence gene in the pathogen. This hypothesis revolutionized the genetics of plant breeding for disease resistance, forcing plant breeders and geneticists to study the evolution and migration of pathogens. Fifty years after its discovery, the gene-for-gene theory still remains valid for understanding the genetics of host-pathogen systems (or pathosystems) (Table **1**).

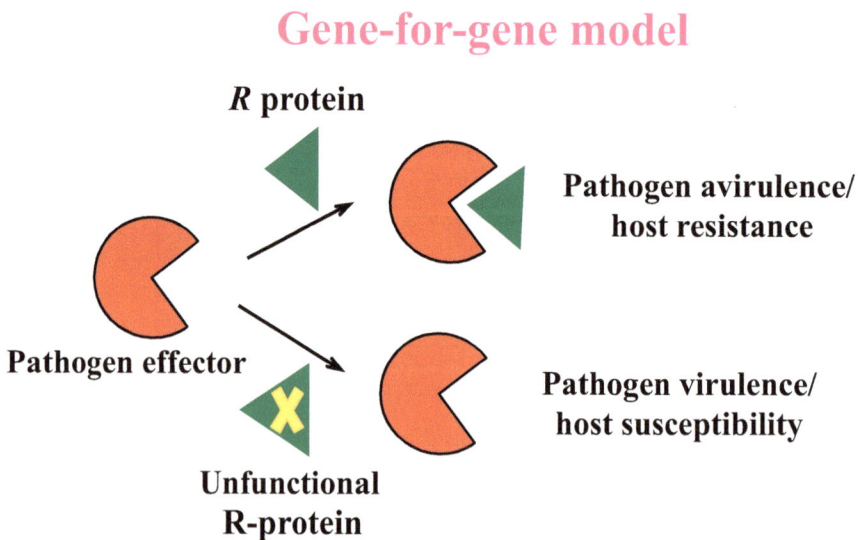

Fig. (1). Mechanism of plant-pathogen recognition according to the gene-for-gene model (modified from Glowacki *et al.* 2011).

Table 1. Host plant and pathogen genotypes leading to compatible or incompatible reactions between plant and pathogen, in gene-for-gene interaction.

		HOST GENOTYPE	
		RR / Rr	*rr*
PATHOGEN GENOTYPE	*Avr Avr / Avr avr*	Avr1 R1 protein Plant and pathogen are incompatible **RESISTANCE**	Avr1 r1 protein Plant and pathogen are compatible **DISEASE**
	avr avr	avr1 R1 protein Plant and pathogen are compatible **DISEASE**	avr1 r1 protein Plant and pathogen are compatible **DISEASE**

The advances made since 1970 in Molecular Biology, have made it possible to better understand the gene-for-gene theory, initially emitted by Flor. In fact, it has been shown that the proteins encoded by resistance genes act as receptors localized at the level of the cell membrane, acting as sensors recognizing in a specific way products (effectors) encoded by the avirulence genes of the pathogen. Whenever these receptors are in contact with their specific ligand, the signal is transferred inside the cell. This signal transduction leads to a hypersensitivity reaction.

2. THE « GUARD » MODEL

This second model (Van Der Biezen and Jones 1998b; Innes 2004) represents an extension of the gene-for-gene model. According to the Guard model, the resistance reaction does not occur simply as a consequence of the direct recognition of the pathogen effector by the resistance protein, but additional cooperation is also necessary between the resistance protein and certain intracellular receptors present in certain hosts, and playing the role of *guarded protein (guardee)* in this tripartite, indirect recognition (Fig. **2**).

Guard model

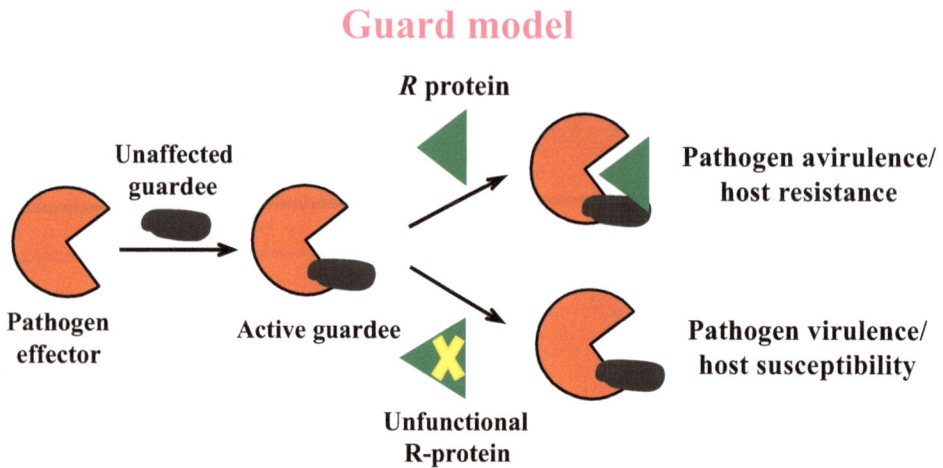

Fig. (2). Mechanism of plant-pathogen recognition according to the Guard model (modified from Glowacki *et al.* 2011).

In *A. thaliana*, the RIN4 protein (RPM1-Interacting Protein-4) is an example of a guardee protein that has been widely studied. This negative regulator of PTI is targeted by the effectors AvrRPM1, AvrPt2, AvrB, AvrPto and HopF2 of *P. syringae* (Mackey *et al.* 2002; Mackey *et al.* 2003; Wilton *et al.* 2010; Deslandes and Rivas 2012). The protein RPM1 (resistance protein) induces defense mechanisms following the perception of RIN4 phosphorylation by the effectors AvrB and AvrRPM1. Similarly, cleavage of RIN4 by the protein AvrPt2 activates ETI through the resistance protein RPS2.

The genome of *A. thaliana* has approximately 167 predicted NBS-LRR proteins (Yu *et al.* 2014). This low number of receptors suggests the importance of the guard model in the recognition of effectors. Indeed, the receptor-ligand model involves the evolution of a large number of receptors in the host plant to be able to specifically recognize each of the effectors secreted by a pathogenic agent during infection. However, the number of pathogens capable of infecting a given species can be very large, further increasing the number of resistance genes necessary for immunity. The guard model, by enabling the activation of defense mechanisms using a single target potentially targeted by several effectors, is therefore a more ***economical strategy*** for plants.

3. THE « DECOY » MODEL

The third model, called the decoy model (Fig. **3**) (van der Hoorn and Kamoun 2008), is a recent modification of the guard model, which argues that specific host proteins, called decoys, associate with the effectors of the pathogen and act as mediators in interactions with resistance proteins.

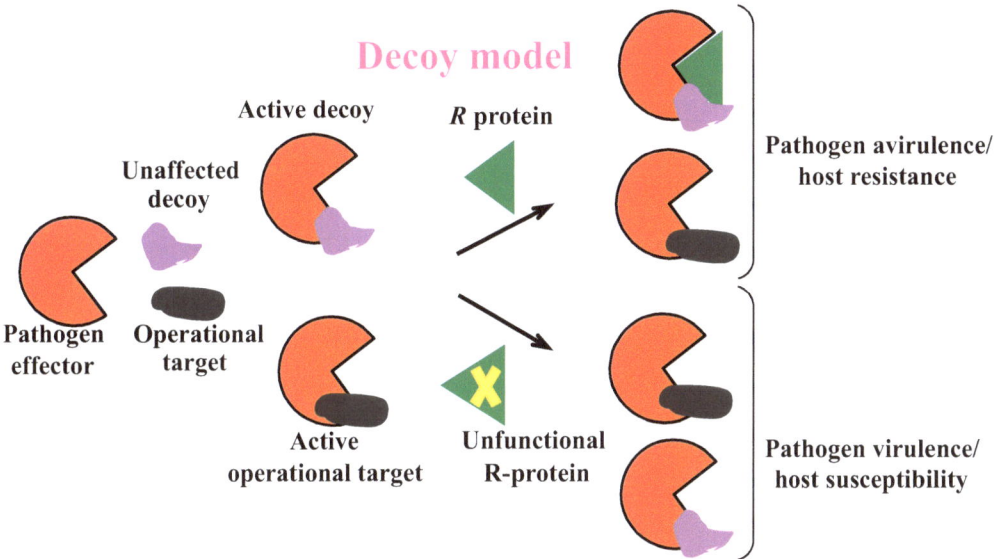

Fig. (3). Mechanism of plant-pathogen recognition according to the Decoy model (modified from Glowacki *et al.* 2011).

The explanation of the decoy model relies on an evolutionary point of view. Actually, the Guard model raises the problem of the double and conflicting natural selection forces that are exerted on the guarded effector target (guardee). This guardee is in an ***unstable evolutionary condition***, which is caused by the polymorphism (presence / absence) of the *R* gene. Actually, in individuals where a functional *R* protein is present, guardees with enhanced pathogen perception should be naturally selected, while in individuals without a functional *R* protein the opposite natural selection force (guardees with decreased effector recognition are favored) is expected to operate.

Thereby, the presence of a decoy is an evolutionary solution that ***relaxes the selection pressure exerted on the guardee.*** Decoy protein is a mimic to the effector target (guardee) and is intended to trap the pathogen into a recognition event. However, the decoy is exclusively specialized in effector perception and has no function either in the development of disease or resistance. Decoys might evolve from effector targets by gene duplication followed by subsequent evolution or evolve independently by mimicking effector targets (target mimicry) (van der Hoorn and Kamoun 2008).

4. INTEGRATED DECOY MODEL (NLR-ID MODEL)

Several recent studies (Cesari *et al*. 2014a ; Ellis 2016; Krattinger and Keller 2016) have shown that NBS proteins (NLRs) with non-canonical domain architectures provide a new molecular model for specific host-pathogen recognition. These composite immune receptors are thought to originate from fusions between NLRs and additional domains that serve as baits for the pathogen-derived effector proteins, allowing pathogen recognition (Sarris *et al*. 2016). Several names were proposed to identify these proteins, such as *integrated decoys*, *integrated sensors* and *NLR-IDs* (Fig. **4**).

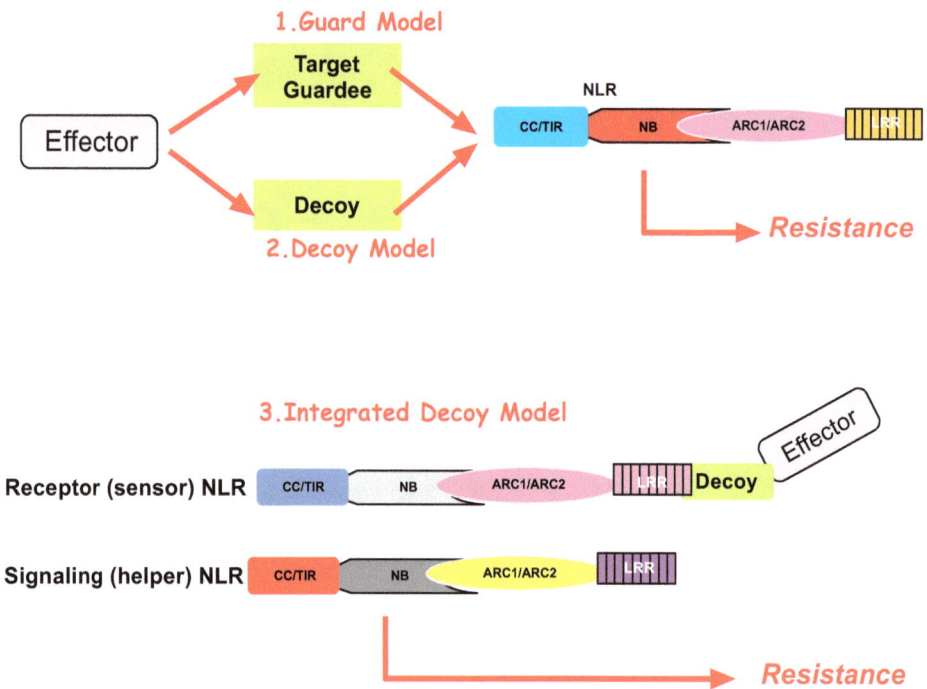

Fig. (4). Model of integrated decoys compared to Guard and decoy models.The decoy is incorporated into the structure of a receptor NLR, allowing effector (Avr) recognition by direct binding (modified from Cesari *et al.* 2014a).

4.1. A Remarkable Diversity of Non-Canonical Integrated Sequences in NLRs

Non-canonical domains have been identified in a large number of functional plant NLRs. Sarris *et al.* (2016) reported the results of a huge and extensive search which has led to the identification of 14,363 NLRs in 40 plant genomes, of which 720 were found to carry 265 distinct integrated Pfam classes. At least 61 of these domains were present in NLRs from more than one plant family, indicating a recurrent theme among integrations. Further, Sarris *et al.* (2016) reported that, on average, 10% of plant NLRs contain IDs. These statistics make quite an impressive set of numbers. The majority of IDs occur as N- or C-terminal fusions with NLRs and the diversity includes integrations of two different domains in the same protein, sometimes at both ends. In a minority, including the rice NLR Pik-1, integration has occurred between the N-terminal signalling domain and the central nucleotide binding domain of the NLR (Cesari *et al.* 2014a).

4.2. Elucidation of the Function of NLR-IDs

The clarification of the function of NLR-IDs is one of the current exciting plant pathology research topics. To date, only some of these non-canonical integrated domains (IDs) have been shown to function in the sensing of pathogen effectors through effector–ID interaction (Nishimura *et al.* 2015). Plant receptors carrying IDs are referred to by Sarris *et al.* (2016) as NLR-IDs, reflecting the fact that, in many cases, the functions of IDs are not yet elucidated. However, for those cases where the ID is related to known or presumed effector target sites in host proteins (virulence targets), the non-canonical NLR domains are referred to as integrated decoys (Ellis 2016).

5. SENSOR NLRS (SNLRS) AND HELPER NLRS (HNLRS)

Recognition of pathogen effectors, on the one hand, and modulation of resistance signaling, on the other hand, is typically thought to be always mediated by *the same* intracellular receptor (NLR). However, recent studies have revealed an increasing number of cases where distinct NLRs cooperate in the recognition of and resistance to pathogens (Eitas and Dangl 2010; Cesari *et al.* 2014a). In most cases, the genes encoding these resistance proteins have neighbour loci in the genome, and are often located in reverse (***head-to-head***) orientation, with a common promoter region, which suggests common expression regulation (Saucet *et al.* 2015; Grund *et al.* 2019). A first case of cooperation was highlighted with the resistance genes *RPP2A* and *RPP2B*, which confer resistance to an isolate (Cala2) of *Peronospora parasitica*[1]. It was concluded that *RPP2A* and *RPP2B* cooperate to specify Cala2 resistance by providing recognition or signaling functions lacked by either partner protein (Sinapidou *et al.* 2004).

In some cases, these NLR pairs seem to act as ***hetero-complexes*** where only one of the paired NLRs acts directly in effector recognition while the other is crucial for the activation of downstream signaling (Williams *et al.* 2014; Cesari *et al.* 2014b) (Fig. **4**) explaining the integrated decoy model). In other cases, several auxiliary (helper) NLRs acting downstream of several sensor NLRs with different recognition specificities have been shown to be necessary for detection and resistance to pathogens (Gabriëls *et al.* 2007; Bonardi *et al.* 2011).

Several effectors of *M. oryzae* are also recognized as pairs of resistance proteins acting in tandem: the ***sensor*** interacts physically with the effector, and the ***helper*** transmits the signal allowing activating the signaling pathways leading to the ETI. Thus, Avr-Pia is recognized by the resistance proteins RGA4 and RGA5, and Avr-Pik is recognized by Pik-1 and Pik-2. RGA4 and RGA5, as well as Pik-1 and Pik-2, are located in reverse orientation in the rice genome (Kanzaki *et al.* 2012; Cesari *et al.* 2013).

Recently, proteins called ***NRC*** (*NLR Required for HR-associated cell death*) have been identified. These NLRs are necessary for triggering ETI and HR, following recognition by several distinct resistance proteins, and these proteins are phylogenetically related sensor CNLs (Wu *et al.* 2017). It was demonstrated that members of the NRC gene family, expanded mainly in Asterids[2] (Lapin *et al.* 2019). It has also been shown that NRC2 and NRC3 present in tomato are required for HR associated with *Prf* (Wu and Kamoun 2019)[3], and that NRC4 is required for cell death associated with *Rpi-blb2*, *Mi-1.2*, and *R1* (Wu *et al.* 2017). In addition, NRC2, NRC3 and NRC4 are redundant and all three contribute to cell death associated with *Rx, Bs2, R8,* and *Sw5* (Wu *et al.* 2017).

The NRC family appears to be an expanding family in *Solanaceae* genomes, and is thought to have evolved from a pair of NLRs acting in tandem. Thus, sensor NLRs (resistance proteins) would have evolved rapidly and would have diversified significantly during evolution, while some helper NLRs would have remained conserved to contribute to cell signaling leading to hypersensitive reaction (HR). This system allows the plant to benefit from broad spectrum resistance while relying on a limited number of helper NLRs (Wu *et al.* 2017).

The significance of the NRC (NLR required for cell death) family is the existence of ***complex NLR immune networks*** that mediate immunity to diverse plant pathogens (oomycetes, bacteria, viruses, nematodes, and insects). Within these complex immune biosystems, helper NLRs are functionally redundant but display distinct specificities toward different sensor NLRs (Wu *et al.* 2017).

CONCLUSION

Understanding how plant and pathogen interact with each other is crucial for developing a sustainable agriculture. The perception of the pathogen is a crucial stage of this interaction, which is performed through a plant sensing system capable of recognizing conserved molecular patterns within specific effector proteins and activating corresponding defenses. Molecular models have been developed and experimentally validated, to explain the basis of pathogen-host recognition. These models have led to a tremendous increase in the understanding of plant-pathogen interactions.

[1] *Peronospora parasitica* (Pers.): Syn. *Hyaloperonospora parasitica* (Pers.) and *Botrytis parasitica* Pers. (source : MyoBank, http://www.mycobank.org/). A phytopathogenic oomycete of the *Peronosporaceae* family. It causes mildew to cruciferous plants such as cabbage and *Arabidopsis thaliana*.

[2] The asterids are one of the largest subgroups of the flowering plants, with more than 75,000 species, including many orders, such as Solanales (Pepper, Tomato, Potato, *etc.*) and Lamiales (*e.g.* olive).

[3] Tomato *Prf* confers resistance against the bacterial speck pathogen *Pseudomonas syringae* pv. Tomato.

CHAPTER 11

PRRs and WAKs: PAMPs and DAMPs Detectors

Abstract: The perception of environmental signals and the ability to react accordingly are essential for the survival of organisms. In plants, extracellular recognition of microbe- and host damage-associated molecular patterns leads to the first layer of inducible defenses, termed pattern-triggered immunity (PTI). Pattern recognition receptors (PRRs) can perceive pathogen/microbe-associated molecular patterns (P/MAMP) from different microbes such as bacteria, fungi, oomycetes or viruses. Danger-associated molecular patterns (DAMPs) correspond to cell wall fragments that can be released by the plant after wounding or pathogen attack. An important group of PRRs is the family of wall associated kinases (WAK) that perceives pathogens indirectly, *via* DAMPs, and activates oligogalacturonide-dependent defense responses. The present chapter will address the most important perception systems used by plants to perceive pathogen attack and initiate efficient defense responses.

Keywords: Bacterial flagellins, Damage Associated Molecular Patterns (DAMP), Pathogen-associated molecular pattern (PAMP), Pattern-Recognition Receptors (PRRs), PAMP-triggered immunity (PTI), Plant innate immunity, Plant lectin receptors, Wall-associated kinases (WAKs).

INTRODUCTION

The immune responses of plants are triggered by the perception of nonself, which can be of a general type or more specific depending on the nature of the recognized molecules. The general-type response (*i.e.* nonspecific) follows the recognition of molecular patterns, common to many microorganisms, called elicitors and more recently PAMPs or MAMPs (pathogen / microbe-associated molecular patterns). When tissues undergo an alteration in their integrity associated with the death of cells, the latter release molecules that are called *danger signals* or damage-associated molecular patterns (DAMPs). Unlike mammals, plants do not have mobile defense cells or an adaptive immune system. Plants contain an innate immune system *embedded in each cell* and emit systemic signals from infected sites (Marina-García *et al*. 2008; Cassel *et al*. 2009; Duewell *et al*. 2010). In all cases, the perception of PAMPs or DAMPs is carried out by receptors called *pattern recognition receptors (PRRs)*, which constitute the first level of recognition of the plant immune system (Jones and Dangl 2006). These receptors confer broad-spectrum resistance because they are able to recognize these conserved molecules that are widely present in microorganisms. These can be receptors with a kinase domain (Receptor-like Kinase: RLK) or protein receptors (receptor like protein: RLP).

In recent years, considerable progress has been made on the functional analysis of PRRs. The first pattern recognition receptor was identified in plants [1] and this led to a ***paradigm change*** in the plant defense field and instigated the recognition of PTI as a critically important component of the plant defense machinery. Since then, several new PRR candidate genes have been identified.

1. PATTERN-RECOGNITION RECEPTORS (PRRS)

1.1. An Overview: Nature of PAMPs and Biochemical Structure of PRRs

Pathogen-associated molecular pattern (PAMP)-triggered immunity (PTI) to microbial infection constitutes an evolutionarily ancient type of immunity that is characteristic of all multicellular eukaryotic living things. Microbial patterns (PAMPs) activating plant PTI have been conserved through evolution ; in other words, these are molecules present in microorganisms, and which are slowly evolving due to the critical functions they exert for survival. These molecules, which are detected by PRRs, are diverse: bacterial (flagellin, elongation factor EF-Tu, and peptidoglycan) (Gust *et al.* 2007), fungal (chitin, xylanase) (Kaku *et al.* 2006), oomycete (β-glucan and elicitins) (Du *et al.* 2015), viral (double stranded RNA) (Niehl *et al.* 2016), and insect (aphid-derived elicitors) (Prince *et al.* 2014).

Plant Pattern-Recognition Receptors (PRRs) mediate microbial pattern sensing and subsequent immune activation. PRRs include a range of non-specific receptors. These receptors often possess leucine-rich repeats (LRRs) that bind to extracellular ligands, transmembrane domains necessary for their localization in the plasma membrane, and cytoplasmic kinase domains for signal transduction through phosphorylation (Zipfel 2014). LRRs are highly divergent, associated with their ability to bind to diverse elicitors. Numerous PRRs rely on the regulatory protein brassinosteroid insensitive 1-associated receptor kinase 1 (BAK1) and other somatic embryogenesis receptor-like kinases (SERKs) (Prince *et al.* 2014)[2].

1.2. Best Known Examples of Bacterial and Fungal PAMPs and their Cognate Pattern Recognition Receptors

The LRR-RLKs FLS2, EFR and Xa21 are capable of detecting bacterial peptides such as flg22 from flagellin, elf18 from EF-Tu and AxYs22 from Ax21, respectively (Kaku *et al.* 2006; Gust *et al.* 2007).

Another example is the plasma membrane LysM receptor-like kinase1 (CERK1/LysM-RLK1), which has been shown to be essential for chitin signaling in *A. thaliana*. Actually, in mutants knocked-out for this gene, chitin-induced resistance against pathogens is suppressed (Miya *et al.* 2007).

Another example is the tomato LeEix1 and LeEix2, which are capable of perceiving the fungal ethylene-inducing xylanase protein EIX (Ron and Avni 2004).

1.3. Focus on FLS2-flg22 Interaction

A very good example is that involving bacterial flagellin, which induces immune responses in many plants (Zhou *et al.* 2010). Mobility due to flagellin is very important for the pathogenicity of bacteria in plants (Schroder and Tschopp 2010). A synthetic peptide of 22 amino acids, flg22, constitutes a conserved domain of flagellin. The plant PRR binding flg22 is a receptor with kinase activity containing an extracellular LRR called FLS2 (Flagellin Sensentive) (Yegutkin 2008). FLS2 engages a cascade of MAPK, WRKY transcription factors and defense effector proteins (Schenk *et al.* 2008) [3]. It is to note that the human counterpart of FLS2, TLR5 (Toll-Like Receptor-5), recognizes different domains of flagellin (Schroder and Tschopp 2010), which suggests a mechanism of *convergent evolution*.

2. A PARTICULAR PRR CLASS: WALL-ASSOCIATED KINASES (WAKS), DAMPS RECEPTORS

Unlike other PRRs that detect pathogenic molecules during infection, other receptors perceive the damage by recognizing the cellular components (DAMPs) that have been disrupted by pathogenic enzymes. Specialized receptors are used by the plant to carry out *indirect recognition of pathogens via DAMPs*. This has been shown in *Arabidopsis* with the perception by WAK1 of oligogalacturonides (Brutus *et al.* 2010) and the perception by DORN1/LecRK-I.9 of extracellular ATPs (Choi *et al.* 2014). WAK1 and WAK2 from *Arabidopsis* perceive oligogalacturonic acid, resulting from the degradation of pectin in the plant cell wall by fungal enzymes (Brutus *et al.* 2010).

2.1. Nature of DAMPs

When the elicitor molecules are of plant origin (generally resulting from the degradation of plant cells), these are ***endogenous elicitors***, which carry molecular patterns called Damage-Associated Molecular Patterns (DAMPs). DAMPs can be peptides (AtPEP1), lipids (cutin monomers resulting from cuticle degradation), saccharides (cellulose oligomers and oligogalacturonic acids (OGA) derived from degradation of the plant wall). Unlike M/PAMPs, which represent nonself, DAMPs are recognized as ***modified self*** (Boller and Felix 2009).

2.2. Example of OGs – WAK1 interaction

Oligogalacturonides (OGs) are the product of the degradation of the plant extracellular matrix by CWDEs (Cell Wall Degrading Enzymes). The most studied CWDEs are endopolygalacturonases (PGs) (ten Have *et al.* 2002). CWDEs are especially important in necrotrophic fungi and fungi having no specialized structure for the penetration of plant tissues (Paccanaro *et al.* 2017). The PGs cleave the bonds between the α1,4-D-galacturonic acid of homogalacturonan, the major component of pectin, thus releasing fragments of OGs (Lotze *et al.* 2007). The latter then play the role of endogenous elicitors making it possible to communicate the damage undergone by the cells to the surrounding tissues or to the whole plant. They are recognized as a DAMP by the Wall Associated Kinase (WAK1) receptor (D'Ovidio *et al.* 2004). WAK1 is part of a large family, WAKs, which in angiosperms make the physical link between the plasma membrane and the cell wall (Anderson *et al.* 2001). The function of AtWAK1 has been identified by Brutus *et al.* (2010) who, with the use of chimeric receptors combining the extra- and intracellular domains of AtWAK1 and other PRRs, showed, on the one hand, that the extracellular domain of AtWAK1 perceived the fragments of OGs, and that, on the other hand, the cytoplasmic kinase domain of WAK1 did induce a defense reaction. AtWAK1 has been shown to be over-expressed in response to OG fragments, while it is under-expressed in the presence of flg22 (Denoux *et al.* 2008).

3. PLANT LECTIN RECEPTORS

Plant lectins are capable of recognizing carbohydrates that originate directly ***from pathogens or from the damage caused during infection*** (Lannoo and Van Damme 2014). Many PAMPs and DAMPs contain carbohydrates (*i.e.* lipopolysaccharides, peptidoglycans, oligogalacturonides, and cellulose) and are recognized by PRRs/WAKs that possess lectin domains, such as lectin receptor kinases (Lannoo and Van Damme 2014).

CONCLUSION

The interaction between receptor and elicitor molecules provides a deeper insight of the molecular process of plant defense. With the development of a clearer picture of the immune response, researchers will be able to create plants with improved resistance to specific diseases and pests. In addition, a more extensive examination of the interaction between elicitor and receptor molecules is needed to distinguish the responses in the cell surface level from cytoplasmic receptor system, and identify the divergences in responses to bacteria, fungi, insects, nematodes, and other biotic components in a particular elicitor–receptor complex.

[1] The first PRR identified in plants or animals was the Xa21 protein, conferring resistance to the Gram-negative bacterial pathogen *Xanthomonas oryzae* pv. *oryzae* (Song *et al.* 1995).

[2] BRI1-Associated receptor Kinase 1 (BAK1) is a member of the Somatic Embryogenesis Receptor Kinase (SERK) family. SERK proteins, which are members of leucine rich-repeat receptor-like kinases (LRR-RLKs), are involved in somatic embryogenesis, an artificial stimulation of natural embryogenesis.

[3] Additional details on FLS2-flg22 interaction are provided in chapter 6 of this book.

CHAPTER 12

NLRs: Detectors of Pathogen Effectors

Abstract: Defense response by NBS-LRR proteins (NLRs) is a sophisticated strategy that induces effector-triggered immunity (ETI). The NBS-LRR proteins are encoded by one of the largest and most important gene families involved in disease resistance in plants. These NBS-LRR proteins are mainly intracellular, and they can specifically recognize effectors secreted by pathogens either directly or indirectly. This will trigger downstream signaling pathways leading to implementation of plant defense response against various classes of pathogens including bacterial, fungal, viral, nematode and insect. In the present chapter we discuss about the present knowledge pertaining to NBS-LRR class of proteins; their structural organization, genomic ditribution and evolution.

Keywords: CNL, Coiled-Coil domain (CC), Gene duplication, LRR (Leucine-Rich Repeats), Nucleotide-Binding Site (NBS), TNL, Toll Interleukin Receptor (TIR).

INTRODUCTION

Cell surface-localized plant innate immune receptors, addressed in the previous chapter, are capable of recognizing diverse evolutionary conserved pathogen components. Conversely, intracellular nucleotide-binding leucine-rich repeat (NBS-LRR or NLR) receptors recognize race-specific pathogen effectors delivered inside host cells. These NBS-LRR receptors are functionally responsible for the detection of a variety of pathogens, including bacteria, viruses, fungi, nematodes, insects and oomycetes. With the advent of high-throughput molecular tools, genomic analyzes have revealed a variable number of putative genes containing an NBS domain, in different cultivated species and models, even if the actual number of functional or expressed *R* genes, in many plant species, remains unknown. Even though NLRs are a stereotyped family of immune receptors, many mechanistic facts remain poorly understood.

1. THE MAIN STRUCTURAL DOMAINS OF NBS-LRR PROTEINS

An NLR tpyically has a modular structure with three domains (N-terminal, central and C-terminal), with distinct roles for each of these three domains. Most *R*

Dhia Bouktila & Yosra Habachi

products would combine a C-terminal receptor domain and a central effector domain (Hammond-Kosack and Jones 1997), which perform two major functions: firstly, the recognition of elicitor molecules by protein-protein interaction mechanisms and, secondly, the direct or indirect activation of transduction signals (Blumwald *et al*. 1998). These signals activate the local hypersensitivity reaction (Fritig *et al*. 1998).

1.1. The C-terminal Region

1.1.1. Leucine-Rich Repeats (LRR) Domain

The C-terminal Leucine-Rich Repeat (LRR) domain is very variable and composed of repeated leucine-rich motifs (and/or other hydrophobic residues). The repeated motif size is 23 amino acids with the consensus (LxxLxxLxxLxLxx(N/C/T)x(x)LxxIPxx) (Jones and Jones 1997). The LRR domains are involved in protein-protein or protein-polysaccharide interactions, as well as in other peptide-ligand associations (Kobe and Kajava 2001). They would therefore carry the specificity of recognition of the avirulence protein (DeYoung and Innes 2006).

A number of proteins with LRR have been described in mammals and *Drosophila*. In these animal proteins, it has been shown that the LRR domain interacts with pathogen-specific proteins (Dangl and Jones 2001). In addition, a great diversity is observed among the different alleles of known resistance genes. In particular the diversity of LRR domains would be at the origin of the diversity of the resistance protein specificities. This assumption is supported by studies carried out on the various alleles of the *L* and *P* genes of flax (Ellis *et al*. 1999; Dodds *et al*. 2000). However, it has been shown that domains of resistance proteins, other than the LRR, also play a role in this specificity (Ellis *et al*. 2000; Luck *et al*. 2000), and that the LRR domain can sometimes have a role in the transduction of the resistance signal (Warren *et al*. 1998; Hwang *et al*. 2000).

1.1.2. Other Domains of the C-Terminal Region

The C-terminal domain in plant NLRs generally contains a Leucine-Rich repeats (LRR) domain; this domain often has detection[1] and self-regulatory functions (Yuen *et al*. 2014). However, the C-terminal region may instead consist of other repeats forming a superstructure (Kobe and Kajava 2000), for example Armadillo (ARM), Ankyrin (ANK), HEAT, Kelch-like repeats, Tetratricopeptide (TPR), WD40, and Pentatricopeptide repeats (PPR) (Sharma and Pandey 2015). Majority of these motifs correspond to repeated units of conserved stretches of 20–40 amino acids, that contain distinctive structures (α-helices or β-sheets) providing them flexibility to bind diverse ligands and proteins.

1.2. The Central NOD Region

This region contains a Nucleotide-Binding Site (NBS) domain with regulatory and oligomerization functions (Inohara and Nunez 2001; Dyrka *et al.* 2014). The central NBS domain corresponds to a site for attachment and hydrolysis of the nucleotide triphosphates ATP and GTP. The analogy of this domain with animal proteins potentially involved in apoptotic cell death phenomena has led to the notion of the NB-ARC domain[2] (Van der Biezen and Jones 1998a). Numerous studies have highlighted the existence of eight major motifs in the NBS domain, some of which are characteristic of one of the CNL or TNL classes and others common to both (Table **1**). Indeed, the four motifs P-loop, kinase-2, RNBS-B (or kinase-3a) and GLPL are common to both classes. However, the two motifs RNBS-A and RNBS-D show no similarity, and the RNBS-C motif has a low similarity between CNL and TNL (Meyers *et al.* 1999). The eighth motif is MHDV, which is highly conserved in the CNL class. The peptide segment containing all of these eight motifs (from P-loop to MHDV) corresponds to the NBS domain, which has a size of around 300 amino acids.

Table 1. Major motifs present in the NBS domain, classified according to their position (Cordero and Skinner 2002; Meyers *et al.* 2003).

S no	Motif	Consensus Aminoacid Sequence	Position Inside NBS Domain
1	P-loop	GVGKTT	1
2	RNBS-A (CNL)	FDLxAWVCVSQxF	20
	RNBS-A (TNL)	FLENIRExSKKHGLEHLQKKLLSKLL	23
3	Kinase-2	LLVLDDVW	74
4	Kinase-3a (synonyme RNBS-B)	GSRIIITTRD	102
5	RNBS-C	YEVxxLSEDEAWELFCKxAF	122
6	GLPL	CGGLPLA	162
7	RNBS-D (CNL)	CFLYCALFPED	223
	RNBS-D (TNL)	FLHIACFF	219
8	MHDV (CNL)	VKMHDVVREMALWIA	~ 300

1.3. N-Terminal Region

1.3.1. TIR (Toll Interleukin Receptor) Domain

This domain has significant sequence homologies with the protein receptor domains isolated from *Drosophila* (Toll receptor) and Man (interleukin-1 receptor) (Hammond-Kosack and Jones 1997). Based on these analogies, a role in the cell signaling cascade is generally assigned to the TIR domain.

1.3.2. Coiled-Coil (CC) Domain

Initially, the Leucine Zipper (LZ) domain is known to have a role in the homo- or hetero-dimerization of proteins (Bent 1996) in the absence of the TIR domain in the N-terminus. Subsequent analyzes of non-TIR-NBS proteins have shown the presence of a Coiled-Coil (CC) domain on the N-terminal region (Pan *et al.* 2000). CC has a primary structure of 7 repeated residues and a tertiary structure of 2 to 5 helices.

1.3.3. Other Domains of the N-terminal Region

NBS and LRR domains are always found in separate proteins in Bacteria, Archaea, protists and algae (Yue *et al.* 2012). However, NBS-LRR genes have been found in two bryophyte species: *Physcomitrella patens* and *Marchantia polymorpha* (Xue *et al.* 2012), and have made it possible to define two new classes of NBS-LRR genes, in addition to the two CNL and TNL. The first class was identified in the *P. patens* genome, and it showed a protein kinase domain at the N-terminus of the NBS-LRR protein defining the ***PK-NBS-LRR (PNL)*** class. The second class was identified in the genome of *M. polymorpha*, and it possessed a hydrolase domain at the N-terminus (***hydrolase-NBS-LRR; HNL***). The CNL class has been found to be divergent from TNL, PNL and HNL in bryophytes (Xue *et al.* 2012).

2. GENOMIC ORGANIZATION OF NBS-LRR LOCI

As in the human major histocompatibility complex (MHC)[3] (Trowsdale 2002), genetic and molecular studies have revealed a particular genomic organization of NBS-LRR resistance genes in plants. Most of them are closely associated with other homologous genes or sequences, together constituting ***complex loci*** or ***clusters*** (Pryor and Ellis 1993). Four types of organization are possible: ***singletons, allelic series, clusters of homologous genes,*** or ***clusters of non-homologous genes.***

The cluster organization corresponds to a set of resistance loci (homologous or not), tandemly organized. All these sequences are not necessarily functional genes (some are truncated products or pseudogenes). Such clustered *R* genes, observed at the genomic scale, could sometimes give birth to ***super- or mega- R genes clusters***. These correspond to groups of genes extending over several million base pairs and containing up to some dozens of genes (Young 2000). For example, the *Arabidopsis* chromosome 6 contains multiple co-localized NBS-LRR sequences, forming a mega-cluster which extends over 4.6 Mb (Meyers *et al.* 1999; Young 2000). The whole-genome sequencing of the model plant *A. thaliana* made it possible to identify 200 NBS-LRR distributed in 40 singletons and 43 clusters (Meyers *et al.* 2003). In lettuce, out of 50 phenotypic resistance genes mapped, 50% (25 genes) colocolize at the three major resistance clusters (MRC1, MRC2 and MRC3) on chromosomes 1, 2 and 4 (Christopoulou *et al.* 2015).

2.1. Simple Locus Organized in Allelic Series

The majority of resistance loci, today listed, have many alleles each conferring resistance to a specific race of the pathogen under consideration. For example, in flax, resistance to *Melampsora lini* is provided by at least 30 alleles grouped into 5 polymorphic loci (*K*, *L*, *M*, *N* and *P*). The *L* locus, in particular, refers to a single gene organized in an allelic series of at least 13 alleles (*L*, *L1* to *L11* and *LH*) corresponding to 13 resistance specificities (Ellis *et al.* 1999; cited in Noir 2002). Similarly, in *A. thaliana*, the locus *Rpp13*, which confers resistance to *P. parasitica*, is a simple locus with 5 distinct functional alleles (Bittner-Eddy *et al.* 2000; cited in Noir 2002). In addition, 10 alleles (*Pm3a* to *Pm3j*) have been assigned to the *Pm3* locus on the short arm of wheat chromosome 1A. All these alleles confer race-specific resistance to powdery mildew (*Blumeria graminis* f. sp. *tritici*). Among them, *Pm3a*, *Pm3b*, *Pm3d*, and *Pm3f* form a true allelic series (Srichumpa *et al.* 2005).

2.2. Complex Clusters of Homologous Resistance Genes

Resistance genes may occur as a series of evolutionarily related (paralogous) genes forming complex loci. Clustered genes, in plant genome, often cosegregate with multiple resistance specificities (Dodds and Rathjen 2010). Some examples include:

- The tomato *(L. esculentum) Cf4/9* loci (Thomas *et al.* 1997).
- The *RPP1* complex locus in *Arabidopsis* (Botella *et al.* 1998).
- and the *Mla* powdery mildew resistance genes of barley *(Hordeum vulgare).*

For example, at the RPP1 complex locus in the *Arabidopsis* accession Wassilewskija, three of four closely related genes, designated *RPP1-WsA*, *RPP1-WsB*, and *RPP1-WsC*, recognize distinct avirulence determinants of the biotrophic oomycete *Peronospora parasitica*, thereby conferring specific resistance to distinct isolates (Botella *et al.* 1998).

On the other hand, up to 30 barley *Mla* variants have been genetically characterized, each of which generates immune responses upon the recognition of the corresponding isolate of powdery mildew biotrophic fungus (*Blumeria graminis* f. sp. hordei) (Shen *et al.* 2003; Seeholzer *et al.* 2010). Cloned genes of the *Mla* locus belong to the coiled-coil, nucleotide binding site, Leucine-rich repeat (CC-NBS-LRR) class of *R* genes (Wei *et al.* 2002). This complex locus involves multiple distinct resistance genes that are tandemly localized within an interval of ~250 kb on barley chromosome 5(1H) (Wei *et al.* 1999). Although evolutionarily related, these genes are quite divergent and, are therefore assigned into three resistance gene homologous families: *RGH1*, *RGH2*, and *RGH3* (Wei *et al.* 1999).

2.3. Complex Clusters of non Homologous Resistance Genes

Unlike **homogeneous clusters** (consisting of an *R* gene and **paralogous** sequences of the same family), **heterogeneous clusters**, rather, include an association of an *R* gene and RGA sequences or other *R* genes from different families. It should be noted, however, that no heterogeneous cluster presenting both a gene of class TIR-NBS-LRR and a gene of class non-TIR-NBS-LRR has been identified (Richly *et al.* 2002).

In multiple plant species, resistance loci, each conferring a specific resistance directed against a different pathogenic agent, are found co-located on the same chromosomal fragment (Jones *et al.* 1993; Witsenboer *et al.* 1995; Spielmeyer *et al.* 1998). For example, at the end of the short arm of the tomato chromosome 6, six resistance specificities are grouped over a region of around 20 cM (cited in Noir 2002):

- two specificities of resistance to *C. fulvum* (*Cf-2* and *Cf-5*).
- Genes *Mi/Meu1*, specifying resistance to the nematode *Meloidogyne incognita* and the aphid *Macrosiphum euphorbiae*[4].
- Gene *Ol-1* resistant to *Oïdium lycopersicum*.
- and gene *Ty-1* governing the tolerance to Tomato Yellow Leaf Curl Virus (TYLCV).

3. EVOLUTION OF THE NBS-LRR GENE FAMILY

The two classes of the NBS-LRR gene family, namely TNL and CNL, are highly diverse and phylogenetically distant with more diversity for the second (CNL) class, among angiosperms (both mono- and dicots) as well as in gymnosperms. Such a wide diversity suggests that duplication of the ancestral gene took place even before separation between angiosperms and gymnosperms (Cannon *et al.* 2002).

A plethora of molecular events (reciprocal recombination, conversions, substitutions, *etc.*) are thought to be at the origin of the complex organization of *R* genes. These mechanisms are an important key to rapid evolution of the *R* genes, and, therefore, are crucial for the host-parasite coevolution process.

In several studies, comparative analyses of the rates of synonyms (Ks) and non-synonyms (Ka) substitutions were carried out within the LRR domains of *R* genes. In most these studies, ***high rates of non-synonymous substitutions*** were observed [5] suggesting that this mechanism is strongly involved in the diversification of LRRs, contributing to their flexible and rapid adaptation potential, as recognitional domains (McDowell *et al.* 1998; Richter and Ronald 2000; Bergelson *et al.* 2001; Ashfield *et al.* 2012; Ruggieri *et al.* 2014; Song *et al.* 2019).

Apart from punctual substitutions, LRRs have also been reported to contain high ***indel*** variation, suggesting that the elasticity of the LRR ***length*** may also contribute to the evolution of the specificity of resistance (Mondragón-Palomino *et al.* 2002). However, genome-wide investigation of members of the nucleotide binding site (NBS)-LRR gene family of *A. thaliana* also revealed substantial evidence of positive selection (a proportion of 30% of positively selected sites) outside LRRs, implying that regions apart from the LRR, also contribute to resistance specificity (Mondragón-Palomino *et al.* 2002).

In addition, the classical process of ***genetic recombination*** (Kover and Caicedo 2001) as well as the phenomena of ***gene conversion*** (unidirectional transfer of genetic information) (Bendahmane *et al.* 2000) constitute a major source of *R* gene diversification. Likewise, ***unequal crossing-over*** events[6], often initiated by the presence of related tandemly repeated sequences, play a key role in the evolution of *R* genes. These unequal recombinations can occur in ***intergenic*** or ***intragenic*** regions. In the first case, they will result in a copy-number variation clustered resistance genes; in the second case, they will create recombinant (potentially active) genes, which could either evolve into new genes, or correspond to pseudogenes. These latter seem to constitute an important evolutionary reservoir of sequences, contributing greatly to the evolution of *R* gene families (Michelmore and Meyers 1998; Hulbert *et al*. 2001; Sekhwal *et al*. 2015).

3.1. The Crucial Role of Duplication in the Evolution of *R* Genes

The number and ratio of unique and duplicated NBS-LRR genes reflects small or large duplication events that have occurred during the evolution of genomes. In *Arabidopsis*, Meyers *et al*. (2003) reported that *local and remote duplications* of TNL and CNL are responsible for the amplification of clusters. It would be important to note that the duplication of genomic regions which contain NBS-LRR genes often corresponds to *functional redundancy*. The presence of genes with redundant functions within duplicate regions has been demonstrated in the soybean genome (Kang *et al*. 2012).

3.2. Diversification of Resistance Genes by Transposable Elements

The abundance of transposable elements (TEs) in eukaryotic genomes and their ability to cause genome restructuring has garnered interest from the research community. By inserting themselves into protein-coding genes, TEs can affect genes (Nekrutenko and Li 2001) either directly through the mutation of the interrupted gene, or *via* the provision of novel regulatory sequences (Jordan *et al*. 2003). TEs can also cause mutation at a larger scale through chromosomal rearrangements such as deletions, duplications, inversions and translocations (Gray 2000).

While TE activity is usually harmful to organism function, some insertions may be of adaptive importance to the host. Thereby, the sequence of a TE can be redirected by the genome and ensure a specific function. This process is called *TE domestication* (Sinzelle *et al*. 2009). TEs are also a source of exons (complete or partial) of several genes coding for functional proteins (Britten 2006). The most plausible mechanism to explain this *exonization* by TEs supposes that these elements will, first of all, be inserted into the introns of genes (where the selective constraint is weaker compared to exons) and that this insertion will be followed by mutations, which will modify the splicing of the gene, thus including part (or all) of the TE in the coding sequence (Nekrutenko and Li 2001). This modification of splicing is facilitated by the presence of several donor and acceptor splicing sites in several TE families. In addition, TEs can also participate in the regulation of genes in their neighborhood. Indeed, Jordan *et al*. (2003) have found that no less than 25% of the gene promoter regions contain elements derived from TEs. These TEs often have, in their sequences, binding sites for various transcription factors (Bourque *et al.* 2008). Furthermore, TEs can participate indirectly in the regulation of genes *via* microRNAs, since it has been shown that several miRNAs derive from TE sequences (Piriyapongsa *et al.* 2007).

Transposable elements play a crucial role in the adaptation of plants to their environment, which includes resistance to abiotic and biotic stress, but also the modulation of pathogens virulence. We can provide the following examples:

a. **Concerning pathogen genomes:** previous studies on TEs in fungal phytopathogens have disclosed their impact on genome plasticity (Pritham *et al.* 2007 ; Haas *et al.* 2009), pathogenicity (Kang *et al.* 2001 ; Zhou *et al.* 2007), host specialization (Yoshida *et al.* 2016) and evolution (Fedoroff 2012 ; Grandaubert *et al.* 2014).

b. **Concerning resistance of plants to abiotic stresses:** Pereira and Ryan (2019) recently found that TEs occurring in the coding sequences or in flanking regions of genes encoding organic anions (OA) transporters could induce mutations, which impact Aluminium (Al) toxicity resistance by modifying the level and/or location of gene expression so that OA efflux from the roots is increased.

c. **Finally, concerning resistance of plants to biotic stresses:** it is noteworthy that there is no evidence for the generation of new specificity at resistance gene loci as a result of the insertion and excision of a transposable element (Richter and Ronald 2000). However, transposable elements have been associated with the generation of variation (mainly through inactivation of *R* genes) in some resistance gene clusters, such as *Hm1*, which confers maize resistance to *Cochliobolus carbonum* race 1, *Xa21*-gene from rice, and the flax *L6* gene. In all these examples, it was demonstrated that the integration of transposable elements into coding sequences disrupted the resistance conferred by the wild allele (Richter and Ronald 2000).

CONCLUSION

Advances in molecular biology have given us much more insight into the plant immune system, and how it can identify the many pathogens plants may encounter. Members of the plant resistance (*R*) proteins family, NLRs, have three distinct domains and are clustered in duplicated genomic regions that are responsible for their amplification. In-depth understanding the molecular function of NLRs (recognition and downstream signaling pathways) greatly depends on advancements in structural and evolutionary studies.

[1]*i.e.* the formation of protein-protein complexes, which can be highly specific,

characterized by high affinity or more flexible.

[2] NB-ARC domain is present in APAF-1 (Apoptotic protease-activating factor1) in Man and the CED4 protein in *Caenorbabditus elegans*. NB-ARC is made of the following initials: NB: Nucleotide-binding; A: *APAF-1*; R: Plant Resistance *(R)* genes; C: *CED4*.

[3] The human MHC is also called the HLA (human leukocyte antigen) complex (often just the HLA).

[4] The aphid-resistance locus, *Meu1*, is tightly linked to the nematode-resistance gene, *Mi*, in tomato. The absence of recombination between the two loci suggests a tight likage or even the same locus (Kaloshian et al. 1995).

[5] This implies that a positive (Darwinian), or relaxed selection are often exerted on the LRR domain enabling its neofunctionalization.

[6] Also referred to as illegitimate recombination. A crossover that occurs between nonequivalent sequences.

CHAPTER 13

Molecular Classification of Plant Resistance Genes

Abstract: Plant resistance (*R*) genes exhibit conserved domains, each of which performs discrete functions in the resistance to pathogens. The most abundant *R* genes belong to the classes of nucleotide binding site leucine rich repeats (NBS-LRR), receptor-like kinases (RLK), and receptor-like proteins (RLP). The list also includes genes encoding proteins with a unique transmembrane domain, genes encoding toxin reductases, genes encoding CC and transmembrane proteins and genes encoding an intracytoplasmic protein kinase. This chapter sheds light on recent advances in the classification of *R* genes, based on their conserved structural characteristics. Knowledge about the R proteins cellular localization and advances in the molecular cloning of *R* genes are also treated.

Keywords: Classes of plant disease resistance (*R*) genes, Cellular localization, Domain architecture, Nucleotide Binding Site-Leucine Rich Repeat (NBS-LRR), *R*-gene cloning, Receptor-Like Kinase (RLK), Receptor-Like Protein (RLP).

INTRODUCTION

Since 1990, the application of molecular approaches such as positional cloning strategy (case of *Pto* in tomato or *RSP2* in *Arabidopsis thaliana*), transposon tagging (case of *N* in tobacco, *L6* in flax and *Cf-9* in tomato) have made it possible, until today, to clone a considerable number of resistance (*R*) genes. A total of 153 reference [1] disease resistance genes have been stored in the latest release of PRGdb database (PRGdb 3.0; www.prgdb.org; Osuna-Cruz *et al.* 2018), in addition to 177 072 putative candidate Pathogen Recognition Genes (PRGs).

Despite the large diversity of parasites to which they confer resistance (fungi, bacteria, viruses, nematodes, *etc.*), sequence comparison reveals the conservation of many structural motifs involved in the recognition and signal transduction. Based on structural protein domains of these *R* genes products and their cellular localization, five major classes, which will be treated in the current chapter, are recognized according to recent literature (Sanseverino *et al.* 2010; Sekhwal *et al.* 2015).

Dhia Bouktila & Yosra Habachi

1. WHY STUDY *R* GENES?

Plant resistance genes are the subject of several research studies not only because of their fundamental relevance to genetic research, but also because of the serious implications that findings in this area of research can have on humans. Plant improvement for resistance to biotic stresses is a major concern, due to the economic losses that pathogens and pests induce in world agricultural production every year. Pathogens and pests limit the yield and performance of crop plants, and threaten food safety at household, national and global scales. Recently, Savary *et al*. (2019) estimated losses in yields due to 137 pathogens and pests associated with wheat, rice, maize, potato and soybean worldwide. Worlwide estimates of yield losses were 21.5% for wheat, 30% for rice, 22.5% for maize, 17.2% for potato and 21.4% for soybean.

Chemical treatment, as a disease and pest controlling method, continues to be widely applied to date, despite its complete inefficiency against certain microbes (viruses, viroids, and mycoplasma) and, above all, its tremendous economic and environmental harmful effects. In this context, plant breeding focusing on the development of cultivars carrying resistance factors to diseases is emerging as an alternative approach that can cancel or minimize costs due to the application of chemicals during agricultural production and, at the same time, reduce their ecological consequences. It is in this context that there is an urgent need to develop our understanding of the immune function of plants. Considering the significance of this problem and the commitment of many scientists, studies in this field are progressing steadily, given the complex and dynamic relationship involving plants and pathogens.

From the point of view of breeders and basic research, the major resistance genes have a number of advantages, such as *efficacy* (at least in the short term), *clear phenotypes* and *simple inheritance*. These advantages have enabled the positional cloning of many *R* genes and almost 20 years after the first successes, we now have many sequences. Knowledge about major *R* genes (including the availability of their sequences) allows to introgress a trait of resistance from one genotype to another by conventional selection or, possibly, by genetic engineering tools.

2. CLASSES OF PLANT DISEASE RESISTANCE GENES BASED ON STRUCTURAL FEATURES

2.1. The Two Classes of Coiled Coil-Nucleotide Binding Site-Leucine Rich Repeat (CNL) and Toll-Interleukin Receptor-Nucleotide Binding Site-Leucine Rich Repeat (TNL)

Proteins of these two classes have a conserved central NBS domain (Nucleotide-

Binding Site), which is the central component of a larger conserved segment called NB-ARC, which is shared by *R* genes in plants, the APAF-1 (Apoptotic protease-activating factor1) in humans and the CED4 protein in the model nematode *Caenorbabditus elegans*[2]. In the C-terminal region, the NB-ARC domain is linked to a variable Leucine-Rich Repeat (LRR) domain, composed of leucine-rich repeated units. A third domain is present in the N-terminal position, which can be either a Coiled-Coil (CC) or a Toll-Interleukin Receptor (TIR) (Pan *et al*. 2000), hence the subdivision into two classes: CC-NBS-LRR (or CNL) and TIR-NBS-LRR (or TNL). For example, the flax *L6* gene that confers resistance to the fungus *Melampsora lini*, encodes a TIR-NBS-LRR (TNL)-type protein (Lawrence *et al*. 1995), whereas the tomato gene *I2*, which provides resistance to race 2 of the ascomycete *Fusarium oxysporum* f. sp. *lycopersici*, the tomato wilt disease agent, encodes a CC-NBS-LRR (CNL)-type protein (Giannakopoulou *et al*. 2015). The NBS-LRR genes are one of the largest families of genes in plants (McHale *et al*. 2006), encoding for cytoplasmic receptors capable of detecting the presence of pathogen avirulence proteins secreted into the cytoplasm[3].

2.2. The two classes of Receptor-Like Protein (RLP) and Receptor-Like Kinase (RLK)[4]

The proteins of the Receptor-Like Proteins (RLP) and Receptor-Like Kinase (RLK) classes have an extracellular LRR domain in the N-terminal region, a transmembrane domain allowing their anchoring to the cell membrane, and a C-terminal domain, which may contain a kinase domain. The RLP class has an intracellular domain, while the RLK class has a C-terminal cytoplasmic Serine/Threonine kinase domain. It is known that the LRR domains, located in the extracellular region, are highly versatile in number of repeated motifs, allowing a very wide range of protein-protein interactions. These include homo- or hetero-dimerization of receptors, in addition to ligand binding.

RLPs are represented, for example, by the family of *Cf* genes conferring race-specific resistance to the biotrophic fungus *Cladosporium fulvum* in tomato (Hammond-Kosack and Jones 1997). The *Xa21* gene is a representative of the RLK class controlling the resistance of rice to *Xanthomonass oryzae* pv. *oryzae* (Song *et al*. 1995).

The **RLK and RLP** R protein classes possess a ***transmembrane domain*** and function primarily as cell-surface sensors for the recognition of pathogenic patterns. It is not the case for the **TNL and CNL** classes, seen above, which usually lack anchoring domains on the membrane, and therefore function mainly as ***cytoplasmic receptors***, directly or indirectly[5] identifying pathogenic molecules, which settle in the host cell after the basal resistance has been defeated.

2.3. Superclass of Oth-*R*-Genes

In addition to these major classes, the PRG database (www.prgdb.org) and recent literature (Sekhwal *et al.* 2015) report the existence of a third superclass grouping together classes, either having none of the structural domains already mentioned or having different associations of these domains (CC-NBS, NBS-LRR, NBS, TIR, TIR-NBS, Kinase, RLK-GNK2). Most of these classes impart resistance by means of several other cellular mechanisms. Thereby, the term oth-*R*, originally introduced by Walter *et al.*, is being used to categorize these unusual *R* genes and *R*-gene analogs (Sanseverino *et al.* 2010; Sekhwal *et al.* 2015).

a. Example of Genes Encoding Toxin Reductases

The *Hm1* gene represents the class of genes encoding toxin reductases; it confers resistance in maize to race 1 of the fungus *Cochliobolus caronum* (Johal and Briggs 1992). It encodes an HC-toxin reductase capable of inactivating the toxin produced by this fungus race, thereby conferring race-specific resistance to corn. These genes are particularly conserved in cereals since counterparts (homologs) have been found in barley or wheat (Han *et al.* 1997).

b. Example of Genes Encoding Proteins With CC Domain and a Transmembrane Domain

A representative of this class of *R* genes is the family of *RPW* genes conferring resistance to *Erysiphe cichoracearum* in *A. thaliana* (Xiao *et al.* 2001). The cytoplasmic proteins RPW8.1 and RPW8.2 are characterized by the presence of a putative N-terminal transmembrane domain, followed by a CC domain involved in interactions with signal molecules (Pan *et al.* 2000). The transmembrane domain, located in the N-terminus, is involved in the direct recognition of an avirulence factor and facilitates the recognition of other molecules (Fluhr 2001).

c. Example of Genes Encoding a Cytoplasmic Protein Kinase

The *Pto* and *PBS1* genes respectively conferring resistance to *Pseudomonas syringae* pv. tomato in tomato (Martin *et al.* 1993) and *P. syringae* in *A. thaliana* (Swiderski and Innes 2001) are two representatives of this class. The Pto protein has similarities to the IRAK and PELLE proteins, protein kinases involved in the immune response in mammals and *Drosophila* respectively (Medzhitov 2001), suggesting common transduction mechanisms between kingdoms.

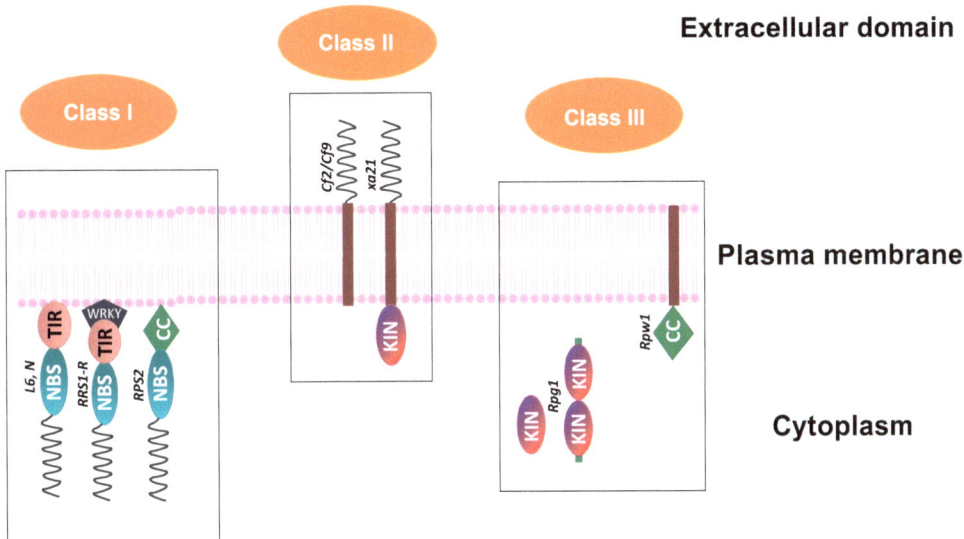

Fig. (1). Major classes of *R* proteins (modified from Joshi and Nayak 2011).

I. The most common *R*-protein family, NBS-LRR, is charcaterized by the presence of a nucleotide-binding site (NBS). NBS-LRRs are located in the cytoplasm as intracellular proteins and exhibit, at their N-terminus, either a TIR domain (*e.g.* L6, N) similar to the animal Toll-interleukin receptor or a predicted coiled-coil region (*e.g.* RPS2). It is possible that an NBS-LRR protein contains an additional domain[6], such as the transcriptional factor WRKY domain (*e.g.* RRS1-R).

II. Members of the RLP family contain extracellular LRRs bound to a transmembrane domain (*e.g.* Cf2/Cf9). The RLK family has an arichitecture made of extracellular LRRs, a transmembrane domain and a cytoplasmic serine-threonine kinase (KIN) domain (*e.g.* Xa21).

III. The third class gathers multiple molecular architectures, for example serine/threonine kinase domain either single (*e.g.* Pto) or double (*e.g.* Rpg1). Also placed in this superclass, RPW8 (*e.g.* Rpw1) gene, which encodes a coiled-coil (CC) motif connected to a membrane anchor.

3. CELLULAR LOCALIZATION OF RESISTANCE PROTEINS

The cellular localization of a resistance gene product can be predicted from its aminoacid sequence. Sometimes these locations are confirmed by wet-lab experiments, such as ***immunolocalization*** (Sauer and Friml 2010). Thus, proteins of the LRR-TM and LRR-TM-kinase types are transmembrane and exhibit their LRR domain in the extracellular space (Piedras *et al.* 2000). Conversely, NBS-LRR proteins are mostly cytoplasmic, although they can rarely have a membrane-spanning region (Boyes *et al.* 1998). Nevertheless, the majority of resistance genes lack signal peptide, and code for receptors (LRR) and transducers (NBS) of protein signals, located inside plant cells, and corresponding to specific avirulence gene products.

Most membrane receptors (PRRs) are of the RLK type (Receptor Like Kinase) (Shiu and Bleecker 2003) with an extracellular LRR region, similar to the Toll type receptors found in animals (Hayashi *et al*. 2001). The best known PRR is FLS2[7]. Nevertheless, PRRs may also have a lectin domain that binds specifically and reversibly to certain carbohydrates, a LysM motif, which fixes peptidoglicanes, or a kinase domain associated with the wall (WAK, Wall-Associated Kinases) (Fig. **2**) (Tör *et al*. 2009). The majority of these receptors have a transmembrane part and a cytosolic part with a kinase domain. Receptor-Like Proteins (Wang *et al*. 2008) have an extracellular LRR domain and a membrane anchor at the C-terminal end but no intracellular kinase domain. The best known of the RLPs is LeEIX1 (Ethylene-Inducing Xylanase), which recognizes xylanase in tomato (Ron and Avni 2004). Finally, PolyGalacturonase-Inhibiting Proteins (PGIP) have only the extracellular LRR domain (Di Matteo *et al*. 2003).

4. POSITIONAL CLONING OF PLANT RESISTANCE (*R*) GENES

Positional Cloning corresponds to the isolation of a gene only from its subchromosomal location, without requiring any information on its biochemical function[8]. The general approach is to define the subchromosomal location of a gene of interest, based on the correlation between genetic recombination frequencies on one hand, and specific molecular markers physically mapped in known regions, on the other hand. Positional cloning is followed by mutation functional analysis, leading to the identification of the molecular origin (*i.e.* the causal mutation) of resistance/susceptibility and the exact biochemical function of the gene.

The availability of a wealth of genetic resources for many plant species [such as whole genome sequence data, high-throughput single nucleotide polymorphisms (SNPs), microsatellites, physical maps, *etc*.] has directly contributed to a significant expansion of the studies on the positional cloning of resistance loci. Cloning novel resistance genes offers the opportunity to transfer the cloned *R* genes into high-yielding varieties by genetic transformation, without compromising their agronomic characters or industrial quality, which may not be easy by conventional, cross-based breeding. Positional cloning provides also, the possibility of incorporating effective resistance genes for heterologous expression in different plant species that cannot be naturally hybridized.

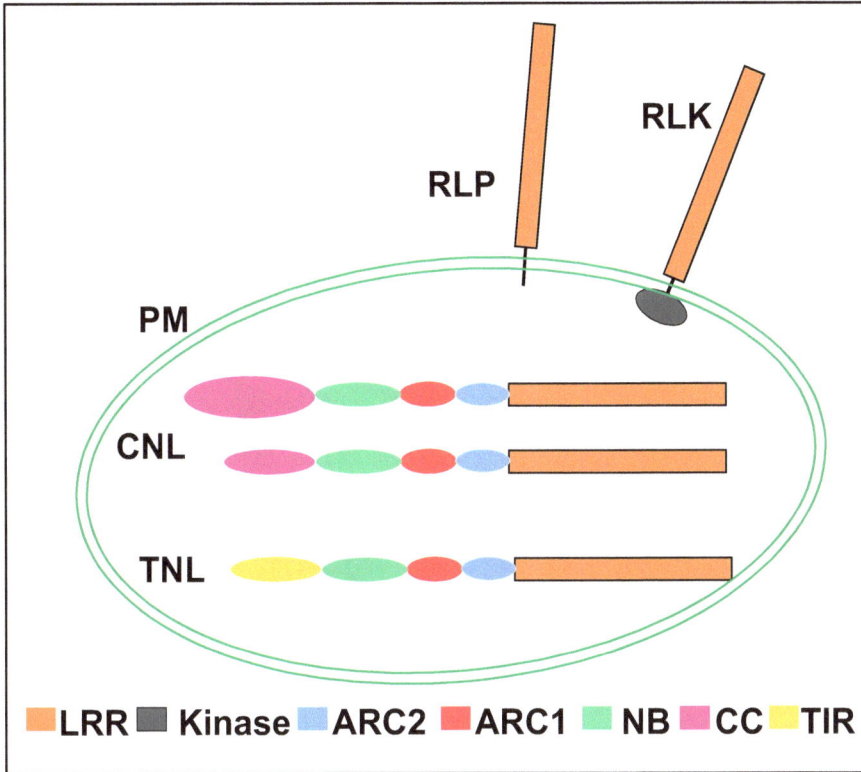

Fig. (2). Organization of extra- and intra-cellular receptor domains in plants (modified from van Ooijen 2007).LRR: leucine-rich repeats; NB: nucleotide binding domain; TIR: Toll-Interleukin-1 Receptor; CC: motif coiled coil; NB-ARC: nucleotide binding domain shared between human APAF-1, (Apoptotic Protease-Activating Factor 1), some plant R proteins, and CED-4 (*Caenorhabditus elegans* Dead protein 4). PM: plasma membrane.

The first-ever cloned *R* gene, namely the maize (*Zea mays*) *Hm1*[9], was published roughly 30 years ago (Johal and Briggs 1992). A year later in 1993, Martin *et al.* (1993) reported the cloning of *Pto* from tomato *(Solanum lycopersicum)*. The year 1994 witnessed a number of publications reporting on the cloning of *R* genes in plants, including RPS2 from *Arabidopsis thaliana* (Bent *et al.* 1994; Mindrinos *et al.* 1994), *Cf-9* from tomato (Jones *et al.* 1994) and *N* from tobacco (*Nicotiana tabacum*) (Whitham *et al.* 1994). Since 1994, many additional cloned *R* genes have been regularly reported. Kourelis and van der Hoorn (2018) recently presented a meta-analysis of 314 cloned functional *R* genes. For these *R* genes, the main classes of encoded products were intracellular NLRs that represented 61% (191/313) followed by cell-surface receptors RLPs/RLKs representing 19% (60/314).

CONCLUSION

Biotic stress is the most significant constraint for plant health and crop yields. Often a single, true-dominant, or semi-dominant resistance (*R*) factor (allele) is needed by the plant to precisely identify infectious and feeding organisms. With the cloning, sequencing and molecular characterization of more and more plant *R* genes, there are now enormous opportunities to understand their functional properties, in terms of host-pathogen interactions and defense signaling pathways capable of ensuring efficient resistance.

[1] (*i.e.* which were validated by experiments in wet labs).

[2] NB-ARC: a nucleotide-binding adaptor shared by APAF-1, certain *R* gene products and CED-4.

[3] See chapter 12 for more detailed information on *R* gene class whose products act as intracellular specific receptors to pathogen effectors.

[4] Proteins from the classes Receptor-Like Proteins (RLP) and Receptor-Like Kinase (RLK) are also named LRR-TM and LRR-TM Kinase, respectively.

[5] Recent advances have shown that certain players, such as the Heat shock protein 90kDa (HSP90), the Suppressor of rps4-RLD 1 (SRFR1), or the Compromised Recognition of TCV (CRT1) could play a role in the activation, accumulation, proper folding and regulation of NBS-LRR proteins and, therefore, are essential for the effector recognition event (reviewed in Elmore *et al.* 2011).

[6] described as non canonical or atypical.

[7] Further detailed in chapters 6 and 11 of this book.

[8] This approach contrasts with the functional cloning, where a prior knowledge of the gene function is required to its cloning.

[9]*Hm1* from maize (*Zea mays*) confers resistance to the ascomycete *Cochliobolus carbonum* Nelson race 1. *Hm1* encodes an enzyme that detoxifies (and, therefore, confers to the plant insensitivity to) the pathogen-derived HC-toxin (Meeley and Walton 1993).

CHAPTER 14

Strategies and Mechanisms for Plant Resistance Protein Function

Abstract: Given the current constraints to sustainable agricultural production, with increasing crop losses due to plant pests and diseases and climate change, considerable advances are required in crop improvement approaches for enabling durable disease resistance. Interestingly, advances in fundamental understanding of the plant immune system will have far reaching implications for genetic resistance development, appropriate for effective and durable disease control and global sustainable agriculture. In particular, a deeper understanding of the molecular and functional mechanisms of resistance (*R*) genes would make it possible to engineer new resistances for future agriculture. In general, there are currently two main strategies, which include nine recognized molecular mechanisms for *R* genes, most of them (all but one, mechanism 6: executor genes) have been used against various types of biotic stress and tend to be widely applicable among plants.

Keywords: Active/passive loss of susceptibility, Direct/Indirect perception, Executor genes, Extracellular/Intracellular perception, Integrated domain, Resistance mechanism.

INTRODUCTION

Resistance (*R*) genes occupy a frontal position in plant immune responses. In innate immunity (with its two branches: specific and nonspecific), the detection of generic or specific molecules emitted by pathogenic attackers is the first stage of the immune reaction. Overall, we can count *nine distinct molecular mechanisms* through which *R* proteins can provide resistance. They are grouped into two general strategies: (I) perception and (II) loss of susceptibility (analyzed in: Kourelis and van der Hoorn 2018).

The *Perception-based strategy* comprises three distinct modes: Extracellular perception (Ia); Intracellular perception (Ib); and Executor genes (Ic). Until date, more than 300 *R* genes have been cloned and characterized, with resistance function confirmed in different plant species (Araújo *et al.* 2019). Most *R* genes encode cell surface or intracellular receptors[1].

Dhia Bouktila & Yosra Habachi

The ***Loss-of-susceptibility strategy*** offers long-lasting resistance, but has a high cost to the host plant and hence can result in reduced yields (Kourelis and van der Hoorn 2018). These authors defined three mechanisms for loss-of-susceptibility: active loss of susceptibility *by **actively disrupting pathogen processes***; passive loss of susceptibility *through **inactivation of pathogen targets** inside the host*; and loss of susceptibility *through **metabolic reprogramming of the host***.

1. STRATEGY (1): PERCEPTION

1.1. Mode (1.1): Extracellular Perception

In this mode, plant cells recognize different pathogen-associated molecular patterns (PAMPs) by plasma membrane-localized RLKs or RLPs. This recognition may be performed ***directly (mechanism 1)*** or ***indirectly (mechanism 2)***.

a. Mechanism 1: Direct Extracellular Perception

Several PAMPs are immediately sensed by RLKs and RLPs on the cell surface. Bacterial flagellin (Felix *et al.* 1999)[2] is the best studied PAMP in plants. In *Arabidopsis*, flagellin epitope flg22 is recognized directly by the receptor kinase (RLK) FLAGELLIN-SENSITIVE2 (FLS2) (Chinchilla *et al.* 2006). The FLS2 gene is ubiquitously expressed among plants (Gómez-Gómez and Boller 2000). However, plants with mutated FLS2 gene exhibit inadequate interaction with flg22, and are, therefore, vulnerable to infection by pathogenic bacteria (Zipfel *et al.* 2004). A number of additional PAMPs are perceived directly, in a way comparable to flagellin. These comprise EF-Tu[3], peptidoglycan, chitin, lipopolysaccharide, and other elements of the bacterial cell wall (Zipfel 2014).

b. Mechanism 2: Indirect Extracellular Perception

The sensing of pathogen molecules, on the cell surface, may also take place indirectly through the detection of altered host factors. A typical case of this mechanism is the perception of the tomato leaf mold, *Cladosporium fulvum* (syn. *Passalora fulva*) by the tomato *Cf-2* protein. The *Cf-2* tomato gene (Dixon *et al.* 1996) encodes an RLP receptor protein with an extracytoplasmic LRR domain that interacts with an extracellular ligand, but lacks a kinase domain[4]. The *Cf-2* gene product activates resistance to strains of the biotrophic fungus *C. fulvum*, which carry avirulence gene *Avr2* (Dixon *et al.* 1996). After being originally thought to have an exclusive receptor affinity to the fungal pathogen, *C. fulvum*,

Cf-2 was also found to mediate resistance to the root parasitic nematode, *Globodera rostochiensis,* because it is capable of recognizing the nematode effector GrVap1 (Lozano-Torres *et al.* 2012). Both Avr2 and GrVAP1 serve as protease inhibitors that cannot be processed by the cysteine protease Rcr3 and directly interact with it (Rooney *et al.* 2005; Lozano-Torres *et al.* 2012). Therefore, Rcr3 is an additional, intermediate, protein that is explicitly needed for *Cf-2*-dependent disease resistance (Luderer *et al.* 2002). This case is one of the rare (if not the unique) examples of an RLP cell surface receptor (*Cf-2*) acting indirectly for Avr2/GrVap1 perception in the apoplast, *via* a guardee/decoy protein, Rcr3 (van der Hoorn and Kamoun 2008).

1.2. Mode (1.2): Intracellular Perception

Plants are not only able to detect pathogen patterns (PAMPs) outside the cell, but also to detect pathogen effectors inside the cell[5]. The majority of cloned *R* genes encode NLRs that are cytoplasmic receptors. In recent years, the central role of the NLR protein family has increasingly been studied in innate immune responses. The NLR protein family, one of the largest multigene families known in plants (Steuernagel *et al.* 2018), may have originated in green algae (Andolfo *et al.* 2019; Gao *et al.* 2018) and existed in early land plant lineages (Yue *et al.* 2012). These NLRs can recognize effectors ***directly (mechanism 3)***, ***indirectly (mechanism 4)***, or ***through integrated domains (IDs) (mechanism 5)***.

c. Mechanism 3: Direct Intracellular Recognition

The direct sensing of effectors is not limited to the cell surface (as explained above), as several effectors are recognized by intracellular NLRs to initiate defense responses. A number of well-known examples illustrating this mechanism are available in literature. Among these, we can cite:

- The direct recognition and physical interaction between RRS1-R and PopP2, an effector of the bacterial wilt, *Ralstonia solanacearum* type III, in the resistant Nd-1 *Arabidopsis thaliana* ecotype (Deslandes *et al.* 2003).
- The direct recognition and physical interaction between the resistance protein Pi-ta and the effector AVR-Pita, during the interaction between rice, *Oryza sativa*, and the rice blast fungus, *Magnaporthe grisea* (jia *et al.* 2000).
- The direct protein interaction involving R proteins encoded by the polymorphic L locus in flax (*Linum usitatissimum*), on one hand, and Avr proteins in various strains of the flax rust fungus (*Melampsora lini*), on the other hand (Dodds *et al.* 2006).

These different associations are a prerequisite to the activation of the defense response in the host (Krasileva *et al.* 2010). In almost all cases of direct physical binding between NLR (receptor) and effector (ligand), the LRR domain has been shown to play a decisive role in the interaction specificity.

d. Mechanism 4: Indirect Intracellular Recognition

Many effectors are recognized by NLRs through an intermediate protein. This occurs either when the effector needs to interact physically with an intermediate protein to be recognized by the NLR, or it modifies enzymatically an intermediate protein to become recognizable by the NLR. Therefore, these additional, intermediate proteins of the host plant often play a pivotal role in the recognition process. These additional host proteins are termed *guardees* or *decoys*[6]. In the guard model, the auxilliary host protein, known as the guardee, is a pathogen effector's defense operative target. In the decoy model, an additional host protein, the decoy, will serve to mimic a true effector target, by attracting the pathogen effector (Van Der Hoorn and Kamoun 2008; Shao *et al.* 2016). Such indirect mechanisms involving guardees or decoys have been evolutionarily selected to allow the host to recognize different pathogens, without the need for different genomically-encoded receptor *R* proteins for every possible pathogen (Dangl and Jones 2001). An illustration of indirect interaction is the perception of the 50-kD replicase fragment (p50) of tobacco mosaic virus by the tobacco NLR N. This interaction involves a third party protein, the N receptor interacting protein 1 (NRIP1), which is a chloroplast-localized rhodanese sulfurtransferase (Caplan *et al.* 2008). N activation is thought to require a prerecognition complex made of NRIP1 and the p50 effector (Caplan *et al.* 2008). An illustration of indirect perception through enzymatic modification of a host protein is the indirect recognition of the *Pseudomonas syringae* Type-III effector HopZ1a by the functionally conserved NLR ZAR1 (Lewis *et al.* 2010). HopZ1a exerts an acetyltransferase activity on the pseudokinase HOPZ-ETI-DEFICIENT1 (ZED1), a nonfunctional receptor-like cytoplasmic kinase (RLCK) (Lewis *et al.* 2013). Therefore, Lewis *et al.* (2013) hypothesized that ZED1 evolved as a decoy to trap HopZ1a to the ZAR1-resistance complex, resulting in effector-triggered immunity (ETI) activation.

e. Mechanism 5: NLR-IDs

Resistance (*R*) genes may also encode NLRs that include additional embedded domains (NLR-IDs), thought to be required for effector recognition (Baggs *et al.* 2017). Integration has been characterized in a number of locations within NLRs, including N- and C-terminal regions, as well as between NLR domains (Cesari *et al.* 2014a; Kroj *et al.* 2016). More than 100 atypical domains within NLR-encoding genes have now been identified across plant genomes such as WRKY and BED domains (Kroj *et al.* 2016; Bettaieb and Bouktila 2020). Analyzing NLR-IDs will offer useful understanding of the actual molecular targets of effectors, since those coupled fragments are probably the segments affected by effectors (Kroj *et al.* 2016). Effector recognition by NLR-IDs may take place either following a direct binding with the pathogen effector or as a result of the effector's enzymatic activity on the NLR-ID (Kourelis and van der Hoorn 2018). NLR-IDs have recently been proven to work with other genomically linked NLRs. Yet, genomically disassembled NLRs may also work cooperatively. An illustration of NLR-ID has been outlined in *Arabidopsis*. Actually, *Arabidopsis* RRS1, which is an NLR integrating a C-terminal WRKY domain, constitutes a receptor complex with RPS4, a helper NLR, to activate effector recognition (Jubic *et al.* 2019). This complex recognizes the bacterial effectors AvrRps4 or PopP2 and triggers the subsequent immune reaction. Both AvrRps4 and PopP2 interact with the RRS1 WRKY domain, and PopP2 functions as an acetyltransferase, acetylating essential lysines in the WRKY domain of RRS1. Upon Avr recognition by RRS1, the complex RRS1/RPS4 becomes activated and RPS4 induces hypersensitive reaction (HR).

1.3. Mode (1.3)

f. Mechanism 6: Executor Genes

Phytopathogenic *Xanthomonas* bacteria secrete transcription-activator like effector (TALE) proteins, into the plant cell where they regulate the transcription of host genes, leading to the host reprogramming and disease establishment. Yet, in certain conditions, the plant can divert these TAL effectors, for its own benefit, to activate immune response to *Xanthomonas* strains containing the TAL effectors (Römer *et al.* 2009)[7]. In the host plant, executor genes can be, therefore, defined as *R* genes that act as a trap for pathogen TALEs, causing them to stimulate transcription of host plant genes associated with the plant immunity. For instance, the expression of rice *Xa27* and pepper *Bs3* resistance (*R*) genes is triggered by the respective TAL effectors AvrXa27 from *X.oryzae* pv. *oryzae* (Xoo) and AvrBs3 from *X. campestris* pv. *vesicatoria* (Xcv) (Römer *et al.* 2009). Recently, the accumulation of knowledge about the operating mode of TALEs has facilitated the production of artificial executor genes controlling immunity toward various *Xanthomonas* strains (Zeng *et al.* 2015).

2. STRATEGY (2): LOSS OF SUSCEPTIBILITY

Several resistance cases depend on the occurrence of mutations in dominant susceptibility alleles, leading to a recessive resistant allele entailing the loss of susceptibility[8] and, therefore, resistance (van Schie and Takken 2014). This strategy is naturally implemented in three manners: active loss of susceptibility (mechanism 7), passive loss of susceptibility (mechanism 8), or loss of susceptibility by host reprogramming (mechanism 9).

g. Mechanism 7: Active Loss of Susceptibility

R genes governing active loss-of-susceptibility express plant proteins that contribute to actively blocking a main function of the pathogen, leading to pathogen neutralization. The mechanisms belonging to this category are flexible and varied; they may work widely against several pathogens, or selectively against a few (Kourelis and van der Hoorn 2018). For example, the *HM1*[9] gene in maize, the first *R* gene cloned (Johal and Briggs 1992), encodes a NADPH-dependent reductase, which specifically inactivates HC toxin, a cyclic tetrapeptide secreted by *C. carbonum* race 1 (CCR1) to promote infection.

Several distinct active loss-of-susceptibility mechanisms of resistance are used against viruses, such as tomato mosaic virus (ToMV) (Ishibashi *et al.* 2014) and Tomato yellow leaf curl virus (TYLCV) Butterbach *et al.* (2014). It was reported that tomato resistance gene *Tm-1* encodes a protein that associates with ToMV replication proteins and, thus, inhibits replication complex formation (Ishibashi *et al.* 2014). On the other hand, the resistance conferred by Ty-1 acts through increasing cytosine methylation of the viral genome[10] (Butterbach *et al.* 2014).

h. Mechanism 8: Passive Loss of Susceptibility due to mutation in a host component targeted by the pathogen

Many of the *R* genes imparting resistance to specific races of the rice bacterial blight disease agent *Xanthomonas oryzae* pv. *oryzae* (*Xoo*) are transmitted recessively (Liu *et al.* 2011), and their mechanism of action tends to require a disruption of the interplay between the pathogen and a specific host component. The dominant allele (*i.e.* the susceptibility allele), *Xa13*, allows bacterial growth *via* the exploitation of the gene by the pathogen effector TALE AvrXa13 (see mechanism 6 above). Interestingly, mutations occurring in the promoter of *Xa13* lead to the resistant, recessive allele, *xa13*, which prevents the gene enslavement, thus resulting in resistance (Chu *et al.* 2006).

i. Mechanism 9: Passive Loss of Susceptibility by Host Reprogramming

Host reprogramming by mutations that distort some plant cellular pathways is a widespread mechanism that results in long-lasting resistance to a wide variety of pathogens (Kourelis and van der Hoorn 2018). This kind of loss of susceptibility is inherited recessively and quantitatively (Ellis *et al.* 2014). The genes involved in this category are often called ***Adult Plant Resistance (APR) genes,*** as resistance typically exists in adult plants (Ellis *et al.* 2014). One example is the recessive alleles (*mlo*) of the barley *Mlo* locus, which gives lesions on leaves and, at the same time, wide-spectrum resistance to powdery mildew. This principle was used to engineer (without genetic transformation) a tomato *mlo* mutant, resistant to powdery mildews (Nekrasov *et al.* 2017).

The nine plant *R* genes' mechanisms of action explained above are summarized in Table **1**.

Table 1. **Plant resistance genes classed according to nine mechanisms of action (from Kourelis and van der Hoorn 2018).**

Mechanism	Description	*R* Genes (Plant Species)
1: RLP/RLK, direct	Recognition triggered by direct interaction of a pathogen-derived effector and a cell surface RLK/RLP receptor.	EFR, FLS2, LORE, LYK3, LYK4, LYK5, LYM1/LYM3, LYM2, RBGP1, RLP23 (Arabidopsis), VvFLS2 (grapevine), NbCORE, NbFLS2 (N. benthamiana), CEBiP, LYP4/LYP6, OsFLS2 (rice), CORE, FLS3, LeEIX2, SlFLS2 (tomato)
2: RLP/RLK, indirect	Recognition triggered either by effector binding to a host component or by effector-mediated modification of a host component, perceived by a cell surface RLK/RLP receptor.	*Cell surface receptor :Cf-2* (tomato) Guardee/decoy: Rcr3
RLP/RLK, unknown mechanism		Ve1 (tomato), StoVe1 (eggplant), HLVe1-2A (hop), NgVe1 (Nicotiana glutinosa), LepR3, RLM2 (oilseed rape), StuVe1, ELR (potato), XA21 (rice), 9DC1, 9DC2, 9DC3, Cf-4, Cf-5, Cf-9, Hcr9-4E, I, I-3 (tomato)
3: NLR, direct	Recognition triggered by direct interaction of a pathogen-derived component and an NLR.	RPP1-{EstA/Nda/ZdrA} (Arabidopsis), L5/L6/L7, M (flax), Roq1 (N. benthamiana), Pi-ta (rice), Sw-5b (tomato)

(Table 1) cont.....

Mechanism	Description	R Genes (Plant Species)
4: NLR, indirect	Recognition triggered either by effector binding to a host component or by effector-mediated modification of a host component, perceived by an NLR.	HRT1, RPM1, RPS2, RPS5, SUMM2, ZAR1 (Arabidopsis), Gpa2, R2, R2-like, Rpi-abpt, Rpi-blb3, Rx1, Rx2 (potato), Rpg1-b, Rpg1r (soybean), N (tobacco), Prf (tomato)
		Guardees/decoys: TIP, RIN4, PBS1, CRCK3, ZED1, ZRK3, RKS1, PBL2 (Arabidopsis), RanGAP2, BSL1 (soybean), GmRIN4 (tobacco), LescPth5, Fen, Pto (tomato)
NLR, unknown mechanism		RBA1, RCY1, RPP13-Nd-1, RPP13-UKID37, RPP13-UKID5, RPP5, RPS6, TAO1 (Arabidopsis), Mla1, Mla10, Mla13 (barley), Bs2 (black pepper), P, P2 (flax), Rxo1 (maize), Fom-2 (melon), L^1, L^{1a}, L^{1c}, L^2, L^{2b}, L^3, L^4, Pvr4, Tsw (pepper), R3a, R3b, R8, Rpi-blb2, Rpi-vnt1.1, Rpi-vnt1.2, Rpi-vnt1.3 (potato), Pi9, Pib, Piz-t (rice), 3gG2 (soybean), N' (tobacco), Bs4, I2, Tm-2, $Tm2^2$ (tomato), Pm2, Pm3a, Pm3f (wheat)
5: NLR-ID	Recognition is triggered either by effector binding to a domain or by effector-mediated modification of a domain that is integrated in a host NLR.	RRS1B, RRS1-R, RRS1-S (Arabidopsis), R1 (potato), Pii-2, Pik-{1/h/p1/s}, RGA5-A, Xa1 (rice)
6: Executor	Recognition triggered by transcriptional activation of the executor gene by a pathogen TAL effector.	Bs3, Bs3-E, Bs4C-R (pepper), Xa10, Xa23, Xa7 (rice)
7: Loss of susceptibility, active	Loss of susceptibility by directly disarming the pathogen by actively interrupting a key pathogenicity process.	JAX1, RTM1, RTM2, RTM3 (Arabidopsis), HvHm1 (barley), Hm1, Hm2, qMdr9.02, qRfg1, ZmTrxh (maize), At1, At2 (melon), IVR (N. benthamiana), PaLAR3 (Norway spruce), STV11-R (rice), Tm-1, Ty-1, Ty-3 (tomato)
8: Loss of susceptibility, passive	Loss of susceptibility by mutation in a host component, leading to the inability to manipulate the host.	rwm1, lov1 (Arabidopsis), Rym-4, Rym-5 (barley), retr01 (cabbage), bc-3 (French bean), Mo-1 (lettuce), sbm1 (pea), $Pvr2^1$, $Pvr2^2$, pvr6 (pepper), Eva1 (potato), xa13, xa25, xa5 (rice), LGS1, Pc (sorghum), Asc-1, Ty-5, Pot-1 (tomato), Tsn1, Snn1 (wheat)
9: Loss of susceptibility, reprogramming	Loss of susceptibility by a deregulated host.	*mlo* (barley), GH3-2, GH3-8, pi21 (rice), Lr34, Lr67, YrL693, Yr36 (wheat)

CONCLUSION

The massive increase in our understanding of plant *R* genes, their products and the molecular interactions responsible for their action opens up immense opportunities both for conventional, biotechnology-, and genomics-based crop improvement, toward a cost-effective, durable and environmentally safe, disease and pest control.

[1] Refer to chapters 11 and 12, respectively.

[2] Further details are provided in Chapters 6 and 11.

[3] Detailed in Chapter 6.

[4] Extensive documentation can be found in Chapter 13.

[5] Extensive documentation can be found in chapters 10 and 12.

[6] Refer to Chapter 10.

[7] Or, as expressed by Tian and Yin (2009): the avirulence TALE proteins betray the pathogen to plant defences during host–pathogen interaction.

[8] Also termed « loss of compatibility ».

[9]*i.e. Helminthosporium carbonum* susceptibility1. *Helminthosporium carbonum* (anamorph: *Cochliobolus carbonum*) is a plant pathogenic ascomycete, with a cosmopolitan distribution. It is a parasite of many *Poaceae* (and secondarily *Rosaceae*) species, including maize (*Zea mays*), sorghum (*Sorghum* spp.), and apple (*Malus domestica*). In maize, it is responsible for a disease called corn-leaf blight disease.

[10] Therefore, as concluded by the authors (Butterbach *et al.* 2014), the results are suggestive of antiviral gene silencing. See the paragraph 1.6. RNA-based antiviral resistance within Chapter 20 of this book, for related documentation.

CHAPTER 15

Signal Transduction Pathways Activated During Plant Resistance to Pathogens

Abstract: The signaling pathways play an indispensable role and act as a connecting link between recognizing the stress molecules and generating an appropriate physiological and biochemical response. Recent studies using genomics and proteomics approach enabled decoding and understanding these signaling networks, which increased our knowledge regarding signaling pathways. This chapter offers a review of the most important bases of plant signaling pathways during various stresses.

Keywords: Abiotic/biotic stress responses, Calcium Signaling, Ethylene, Jasmonic acid (JA), MAPK cascades, Plant signaling pathways, Phytohormone signalling, ROS (Reactive Oxygen Species), Salicylic acid (SA), Signaling cross talk.

INTRODUCTION

Pathogen recognition leads to the execution of intracellular signaling events that contribute to the activation of of the plant's adaptive response (Jones and Dangl 2006). These events may involve a large number of actors. Among the first signaling events detecetd following the perception of a pathogen are *ion flows* through the plasma membrane, notably *influx of Ca^{2+}* and *effluxes of NO^{3-} or Cl then K^+*. These ion movements trigger a membrane depolarization; and the amplitude and duration of this depolarization depends on the elicitor (Garcia-Brugger *et al.* 2006).

These studies have been, more recently, complemented by the discovery of the participation of certain phytohormones (initially known for their involvement in development) in the implementation of adaptive responses of plants to biotic and abiotic stresses (Pieterse *et al.* 2009). This is the case for *brassinosteroids*, *auxins (AUX)*, *gibberellins (GAs)*, *cytokinins (CK)* and *abscissic acid (ABA)*. These various hormones play a pivotal role in regulating the signaling network of the plant defense system. However, the different pathways specific to each of these

Dhia Bouktila & Yosra Habachi

hormones can be interdependent and, thus, cross-communicated in an antagonistic or synergistic manner, giving the plant with a powerful ability to finely regulate its immune response (Pieterse *et al.* 2009).

1. PHYTOHORMONE SIGNALING

In addition to their involvement in many physiological processes, phytohormones are signaling molecules that regulate plant immunity (Pieterse *et al.* 2012). All phytohormones, including salicylic acid (SA), jasmonic acid (JA), ethylene, abscissic acid (ABA), giberillin (GA), auxin (Aux), cytokinin (CK) and brassinosteroid (BR) participate in plant immunity and form complex signaling networks to coordinate responses to various stresses. Among them, salicylic acid (SA), jasmonic acid (JA) and ethylene (ET) are the main immune phytohormones, whose major role in resistance to pathogenic microorganisms has been demonstrated since many years (Glazebrook 2005; Lorenzo and Solano 2005; Broekaert *et al.* 2006; Loake and Grant 2007; Balbi and Devoto 2008).

1.1. Salicylic Acid (SA)

Studies carried out on the physiological role of salycilic acid (SA) have demonstrated that this compound has a key role in the establishment of basal resistance (innate immunity), in the execution of RH, and in the establishment of SAR. A large number of mutants or transgenic plants relating to this defense pathway have been characterized. The biosynthesis of SA can be initiated in plants at the level of chloroplasts where two enzymatic pathways using *chorismate* as a precursor have been identified (Vlot *et al.* 2009). Studies on tobacco suggested that SA is derived from the phenylpropanoid pathway initiated by phenylalanine ammonia lyase (PAL); it would be synthesized from trans-cinnamic acid and benzoic acid (Shah 2003) in the cytoplasm. Other work has shown that the isochorismate pathway is a major source of SA during SAR in *A. thaliana* (Wildermuth *et al.* 2001).

1.2. Jasmonic Acid (JA) and Ethylene

Two other hormones play an important role in hormone signaling pathways: jasmonic acid (JA) and ethylene (ET). These two hormones often act synergistically when building resistance. The three main components of the JA pathway are coronatin insensitive 1 (COI1), jasmonate resistant 1 (JAR1) and Jasmonate ZIM Domain 1 (JAZ1) (Fonseca *et al.* 2009). COI1 is a protein involved in the degradation of proteins *via* the proteasome 26S, and is necessary for almost all the responses implemented by the JA. JAR1 encodes a JA amino acid synthetase involved in the synthesis of isoleucine-JA (Ile-JA), a bioactive molecule that can diffuse from cell to cell and induce the expression of genes associated with resistance to necrotrophic pathogens and insects (Staswick and Tiryaki 2004). Ile-JA promotes interaction between COI1 and JAZ1. Finally, JAZ1 is a repressor of the transcription of genes responding to JA (Thines *et al.* 2007), and the interaction between COI1 and JAZ1 leads to the degradation of JAZ1 and to the removal of the inhibition of defense genes.

Ethylene is an important factor in responses to various stresses in plants, such as mechanical injury and infection by pathogens (Guo and Ecker 2004).

2. CALCIUM SIGNALING

Calcium plays a key role in the growth and development of plants. Ca^{2+} signaling has been implicated in various pathways and responds to many extracellular stimuli, such as light, abiotic and biotic stress factors, all causing cellular calcium levels to change.

Two opposing reactions may occur within the plant cell: Ca^{2+} influx through channels or Ca^{2+} efflux through pumps. Actually, the removal of Ca^{2+} from the cytosol against its electrochemical gradient to either the apoplast or the intracellular organelles requires energized 'active' transport. The high level of Ca^{2+} can be recognized by calcium sensors or Ca^{2+} binding proteins, which together can activate protein kinases (Mahajan *et al.* 2006). These activated protein kinases can phosphorylate many regulatory proteins, including transcription factors, which regulate the level of gene expression, resulting in a change in metabolism, followed by a physiological response to stress resistance/tolerance. The physiological response may consist of an inhibition of plant growth or cell death, which will depend on the number and type of genes that are up- or down-regulated in response to high calcium (Tuteja and Mahajan 2007).

Several proteins exist in plants that bind to Ca^{2+} to modulate several processes. For example, we can cite the ***Calcineurin B-like proteins (CBL)*** in *Arabidopsis*, which form a complex with the CIPKs (Albrecht *et al.* 2001; D'Angelo *et al.* 2006). It has been proposed that this property of the CBL/CIPK complex contributes to efficient signal transduction.

Another calcium sensor in plants is ***Calmodulin (CaM)***. It is a highly conserved Ca^{2+} sensor, devoid of intrinsic enzymatic activity and functioning by regulating the activities of its targets (CaM-binding proteins: CaMBP). The accumulated data, in *Arabidopsis* and other organisms, indicate that Ca^{2+}/CaM is involved in the regulation (or reprogramming) of transcription [1], by binding to various transcription factors, in particular the bHLH factor (Corneliussen *et al.* 1994) and the family of transcription factors AtBT (Du and Poovaiah 2004), thus allowing the plant to regulate cell functioning during development or stress response (Bürstenbinder *et al.* 2017). Once linked to these transcription factors, Ca^{2+}/CaM can activate or inhibit their DNA binding capacity and therefore, influence the transcription of the target gene downstream (Snedden and Fromm 2001). Park *et al.* (2005) revealed that WRKY transcription factors of group IId of *A. thaliana* interacted with calmodulin *via* a link segment (VAVNSFKKVISLLGRSR) at the N-terminal end of the WRKY factor.

3. MAPK CASCADES

By regulating Mitogen-activated protein kinase (MAPK) cascades, plant cells are able to respond to a range of abiotic and biotic constraints including temperature, UV rays, ozone, reactive oxygen species, drought, osmolarity, heavy metals, injuries and pathogen infections (Sinha *et al.* 2011; Opdenakker *et al.* 2012; Danquah *et al.* 2014; de Zelicourt *et al.* 2016). With regard to the response to biotic stress in particular, it has been shown that MAPKs regulate various physiological defense mechanisms, such as the synthesis of phytoalexins[2] or the reprogramming of gene expression[3] (Danquah *et al.* 2015).

MAPK cascades are ***evolutionarily conserved*** in all eukaryotes, including plants, fungi and mammals (Zanke *et al.* 1996; Ligterink and Hirt 2001; Xu *et al.* 2017). MAPKs, which are present both in the cell cytoplasm and in the nucleus, therefore make it possible to convert extracellular stimuli into intracellular responses. Indeed, following the perception of a stress signal, the activation of the cascade generally begins with a MAP3K kinase (MAPKKK), which activates by phosphorylation a second MAP2K (MAPKK), which, in turn, phosphorylates and activates its MAPK downstream. MAP3Ks are serine/threonine kinases, which phosphorylate their MAP2Ks downstream, doubly on the residues of serine and threonine. MAP2Ks are threonines/tyrosine kinases and phosphorylate their MAPKs downstream on TEY or TDY motifs [4] of their activation loop (Jonak *et al.* 1994). Similar to animals and yeast (*S. cerevisiae*), plant MAPK are encoded by multigene families (Sinha *et al.* 2011).

4. THE OXYDATIVE BURST

Reactive Oxygen Species (ROS), or Reactive Oxygen Intermediates (ROI), are compounds derived from oxygen, produced throughout the life of the plant. Some are toxic byproducts[5] from plant metabolism, but they can also be produced in response to biotic and abiotic stress. In the case of the response to pathogens,

ROS play the role of antimicrobial compounds, participating in the strengthening of the wall and are also considered as signal molecules (Gechev and Hille 2005). The most abundant forms of ROS are the superoxidic ions O^{2-} and Hydrogen peroxide H_2O_2.

During the interaction with the majority of pathogens, a biphasic oxidative burst is observed: a first, very rapid, not pathogen-specific peak of H_2O_2, will be followed by a second prolonged peak, observed only during incompatible interaction (Wojtaszek 1997). It causes cell death due to the high toxicity of the molecules produced. In order to adjust the levels of ROS in the cell, the plant has an arsenal of genes involved in the regulation of these species. Indeed, it has been reported that at least 152 genes are involved in the production and detoxification of ROS in *Arabidopsis* (Mittler *et al.* 2004). The multiple ROS production sites, generated in all cell compartments, highlight the complexity of this regulatory network. The main pathway of ROS production involves NADPH oxidases, which are present on the membranes of chloroplasts, mitochondria and peroxisomes (Torres and Dangl 2005).

Even if ROS (which are highly toxic) cause some damage to the cell, the oxidative burst is essential to trigger a cascade of events leading, ultimately, to the activation of defense genes. Antioxidants play a crucial role due to their ability to neutralize certain toxic oxygen derivatives. In addition, the interactions of ROS with other signaling molecules such as nitrogen monoxide (NO), lipids or even plant hormones, also participate in this signaling (Gechev and Hille 2005). The ROS will diffuse into the cytosol and activate the MAPK/CDPK pathways as well as several defense responses such as the production of phytoalexins and glucosinolates and lead to HR-type cell death to limit the spread of the pathogen (Fig. **1**).

5. MAIN PATHWAYS TRIGGERED DURING RESISTANCE TO BACTERIA

Induced Systemic Resistance (ISR) [6] is an acquired systemic resistance, which is triggered following the colonization of the plant by certain strains of non-pathogenic bacteria (Pieterse *et al.* 2014). ISR is dependent on the ET and JA pathways and is generally associated with resistance against necrotrophic pathogens (Shen *et al.* 2018).

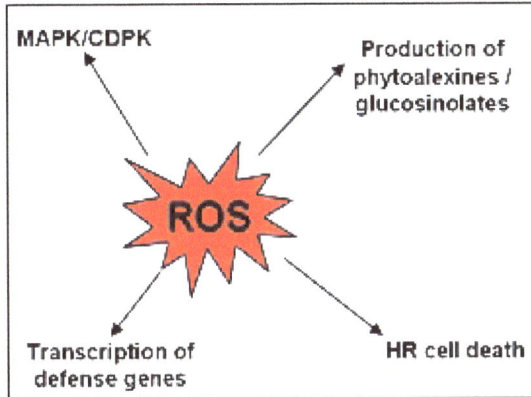

Fig. (1). Roles of reactive oxygen species (ROS) during plant/pathogen interactions.

6. MAIN PATHWAYS TRIGGERED DURING RESISTANCE TO BIOTROPHIC FUNGI

Signaling dependent on SA production is generally associated with the attack of biotrophic or hemibiotrophic pathogens. ET has complex roles that differ depending on the pathosystem. For example, the ethylene-insensitive mutants of *Arabidopsis* (*ein2*) are more resistant to certain biotrophic pathogens but they are more sensitive to certain necrotrophic pathogens (Shibuya *et al.* 2004; Robert-Seilaniantz *et al.* 2007).

Indole 3-acetic acid (IAA, 3-AIA) not only performs many functions in plant growth, but also in response to pathogens. Little is known about the mechanisms of action of this hormone in biotic stresses. In this context, the signaling pathways of the IAA act mainly in an antagonistic manner with those of the SA (Fu *et al.* 2011). Indeed, the activation of pathways associated with IAA is correlated with an increased sensitivity towards biotrophic fungi as well as with a decrease in the biosynthesis of SA, because IAA acts antagonistically to SA necessary to biotrophic resistance. For instance, the exogenous supply of auxin suppresses the expression of *PR1* (*pathogenesis regulated 1*), a gene strongly induced by SA (Robert-Seilaniantz *et al.* 2011).

7. MAIN PATHWAYS TRIGGERED DURING RESISTANCE TO NECROTROPHIC FUNGI

JA- or ET-dependent signaling is associated with defense responses induced against necrotrophic pathogens (Zimmerli *et al.* 2004; Glazebrook 2005). Abscissic acid (ABA) can positively influence the development of resistance against necrotrophic pathogens by acting in concert with JA (Mengiste *et al.*

2003). Other studies have shown that ABA can also interfere with ET pathways (Cheng *et al*. 2009). The insensitivity to ET and/or the accumulation of ABA makes plants more sensitive to certain pathogens such as the fungus *Fusarium oxysporum* (Berrocal-Lobo and Molina 2004).

8. SIGNALING CROSSTALK BETWEEN PLANT ABIOTIC AND BIOTIC STRESS RESPONSES

The relationship between disease resistance in plants and environmental conditions has been developed by Schoeneweiss (1975). Indeed, when abiotic stress is exerted on a host plant, even before infection with biotic stress, this will increase the sensitivity of this plant to the pathogen. It becomes evident that the response of plants to their simultaneous or sequential exposure to **combined stress** is distinct from the response to **individual stress** (Zhang and Sonnewald 2017; Vemanna *et al*. 2019).

It is currently recognized that there is a multi-level, interconnected signaling network. During the **crosstalk**, there is a transfer of important information between the signaling pathways of the response to various abiotic and biotic stresses, using **common components**.

The outcome of combined stress (for example, plant-pathogen interactions under abiotic stress) depends on many factors, including the genotype of the plant, its age, the genotype of the pathogen, the mode of infection, as well as the nature, strength and timing of application of abiotic stress (Bostock *et al*. 2014). In general, abiotic stresses increase sensitivity to hemibiotrophic or necrotrophic pathogens (including facultative and weakly virulent pathogens), but reduce sensitivity to biotrophic pathogens (Saijo and Loo 2020).

Common signaling modules are generally involved in early responses under various biotic/abiotic stress conditions, before the branching of the stress-specific signal (Zhang and Sonnewald 2017). It is known that salt stress often coincides with other abiotic/biotic stresses (Mantri *et al*. 2010; Bai *et al*. 2018). For example, in tomato, significant interactions have been observed between resistance to powdery mildew and the severity of salt stress. Indeed, the accumulation of Na^+ and Cl^- in the leaves negatively influences the growth of powdery mildew (Bai *et al*. 2018). Further, salt stress significantly increased the susceptibility of tomato plants to powdery mildew (Kissoudis *et al*. 2015). Another example, combined heat and drought stress increased the susceptibility of *Arabidopsis* plants to Turnip mosaic virus infections through suppression of defense responses to the biotic stress (Prasch and Sonnewald 2013). Moreover, the study of the molecular response of *A. thaliana* to co-occurring drought stress and plant-parasitic nematode, *Heterodera schachtii*, infection using microarrays revealed that drought stress increased the susceptibility of *A. thaliana* to nematode infection (Atkinson *et al*. 2011). Furthermore, induction patterns of differentially expressed genes following drought (abiotic) and nematode (biotic) stress treatments, in *Arabidopsis*, displayed not only a specific response to each stress but also a particular response that was uniquely activated by the combination of biotic/ abiotic stress (Atkinson *et al*. 2011). On the other hand, Prasch and Sonnewald investigated the response of Arabidopsis plants to heat, drought, and Turnip mosaic virus (TuMV) by transcriptome and metabolome analysis. Their results revealed that the highest number of TIR-NBS-LRR resistance genes is induced under heat and combined heat and drought stress, whilst only a few TIR-NBS-LRR genes were regulated under combined triple stress. These findings suggest that HSPs regulate the function of NBS-LRR genes in response to biotic and abiotic stresses (Prasch and Sonnewald 2013).

CONCLUSION

Understanding complexity of stress signaling is indispensable in order to minimize the impact of stresses on plant growth, development, and productivity and maximize their lifespan. The major elements of transduction pathways, described in this chapter, are only a small part of the vast signaling network leading to resistance, and these pathways are not exclusive. In addition, there are resistances to certain pathogens that are not associated with any of these pathways. Further, understanding signaling crosstalk between components of the plant defense pathways under combined stress conditions is still limited, although it is continually evolving due to remarkable advances in high-throughput sequencing technologies and powerful bioinformatics tools for genome-wide study, transcriptome, and proteome analysis.

[1] See chapter 16: Transcriptional Reprogramming in Plant Defense.

[2] Refer to chapter 18: Plant Defense Gene Expression and Physiological Response.

[3] Refer to chapter 16: Transcriptional Reprogramming in Plant Defense.

[4] T= Threonine; E= Glutamic Acid; D= Aspartic Acid; Y= Tyrosine.

[5] *i.e.* secondary product derived from a production process.

[6] Refer to paragraph 2.2. Acquired resistance, in Chapter 4.

Transcriptional Reprogramming in Plant Defense

Abstract: In order to set up an optimal response to the invading pathogen, the host plant uses transcriptional reprogramming. This phenomenon, involving both DNA-binding transcription factors (TFs) and their regulatory molecules, occurs at several levels of resistance, such as the expression of resistance components (*e.g.* intracellular and membrane receptor proteins), and downstream defense signalling. In this chapter, we address the structure and function of main TF families associated with plant defense against biotic stress, as well as the role played by MAPK cascades and Ca^{2+} signaling in regulating transcriptional complexes during plant-pathogen interactions.

Keywords: BZIP, ERF/DREB, MYB, NAC, Pathogen-triggered cellular responses, Signalling network, Transcription factors, WRKY.

INTRODUCTION

In multicellular organisms, the expression of genes coding for proteins is subject to complex regulation in space (tissue) and time (developmental stage). Such specificity of gene expression is partially attributable to the action of proteins capable of activating or repressing transcription: *transcription factors (TFs)*. They are generally composed of a DNA-binding domain, a nuclear localization signal (NLS) and a transcription activation domain allowing them to modulate the level of transcription of their target genes. The transcription factor (TF) genes represent an important group targeted for crop improvement. These factors play an active role in the initiation of the transcription process as well as the control of development and response to external stress throughout the life cycle of an organism. They are able to induce a cascade of metabolic pathways, by modifying the transcription profile of a plant cell. These metabolic adjustments help plants react to harsh environmental conditions, since plants are sessile and cannot escape environmental stress.

In particular, several transcription factors are critical players in stress resistance and plant adaptation to external constraints, because they govern the transcription of almost all stress-sensitive genes, by binding to their cis elements.

Dhia Bouktila & Yosra Habachi

1. MAJOR TRANSCRIPTION FACTOR FAMILIES ACTIVE IN PLANT IMMUNITY

Numerous TF families, including WRKY, MYB, NAC, and bZIP (Table **1**), are involved in stress response, and the expression of their coding genes is associated with the enhancement of resistance/tolerance in both crop and model plant systems (Wang *et al.* 2016a; Baillo *et al.* 2019).

Table 1. Summary features of the discussed transcription factor (TF) families.

TF Family	DNA-Binding Domain	*Cis*-acting Element	Structural Features
WRKY	WRKYGQK domain	W-box (TTGACT/C)	• **N-terminus**: WRKY domain of ~60 amino acid residues. • **C-terminus**: Zinc-finger structure ($Cx_{4-5}Cx_{22-23}HxH$ or $Cx_7Cx_{23}HxC$).
NAC	NAC domain	NACRS (TCNACACGCATGT)	• **N-terminus**: NAC domain of 150 amino acids residues. • **C-terminus**: Variable transcription regulatory.
MYB	MYB domain	MYBR (TAACNA/G)	Multiple repeats each of about 52 amino acids, forming a helix–turn–helix (HTH) structure.
ERF/DREB	AP2/ERF domain	GCC box (AGCCGCC) and (TACCGACAT)	A conserved domain of 50-70 aminoacids consisting of three parallel β-sheets and an α-helix.
bZIP	bZIP domain	C-box (GACGTC), A-box (TACGTA), G-box (CACGTG), PB-like (TGAAAA), and GLM (GTGAGTCAT)	A sequence of 60–80 amino acids divided into :1) a conserved 18-amino acids basic region N-x7-R-K-x9 responsible of DNA-binding; and 2) a leucine zipper.

1.1. WRKY Transcription Factors

The family of WRKY transcription factors is distinguished by a strongly conserved 60 amino acid sequence denoted the WRKY domain, which contains a standard WRKYGQK motif, supplemented by a zinc finger motif Cys2His2 or Cys2HisCys (Eulgem *et al.* 2000). WRKY proteins bind specifically to the W-Box element [(T)TGCA(C/T)] (Du and Chen 2000). Seventy-four WRKY transcription factors were reported in *Arabidopsis* (Eulgem *et al.* 2000).

WRKY transcription factors are among the major groups of regulators of gene transcription in plants and are an essential part of the signaling pathways that govern multiple physiological functions in plants. New discoveries show that

WRKY TFs have a role in regulating important plant functions. In addition, the regulation of many apparently different processes may be provided by a single WRKY transcription factor (Lee *et al.* 2018). Signaling and transcriptional regulation mechanisms have been dissected, discovering the interaction of the WRKY with multiple protein collaborators, such as *mitogen-activated protein kinase* (MAPK or MAP kinase), MAPKKs, calmodulin, Histone deacetylase, resistance proteins as well as other WRKY transcription factors (Rushton *et al.* 2010; Banerjee and Roychoudhury 2015).

WRKY factors have been studied in the gene response to diseases in several species such as wheat (Gupta *et al.* 2019), corn (Wei *et al.* 2012b), chickpea (Waqas *et al.* 2019) and rice (Ramamoorthy *et al.* 2008). WRKYs are also involved in the response to abiotic stress (Marè *et al.* 2004). In particular, the factor WRKY70 seems to play a role in the interconnection between SA and JA signaling pathways.

Furthermore, WRKY TFs can form a convergence point in cross-talk of ovelapped pathways of abiotic and biotic stress response. For instance, Lee *et al.* (2018) demonstrated that rice OsWRKY11 can stimulate drought tolerance and resistance to the bacterial pathogen, *Xanthomonas oryzae* pv. *oryzae*, through positively regulating both abiotic and biotic stress-responsive genes.

1.2. NAC Transcription Factors

The NAC family is one of the largest families of plant-specific transcription factors, and the expression of its members is modulated during the development of the plant and its response to biotic and abiotic stresses (Olsen *et al.* 2005). The acronym NAC is derived from NAM (No Apical Meristem), ATAF (*Arabidopsis* Transcription Activation Factor) and CUC2 (Cup-shaped Cotyledon), which are genes that are independently characterized and all contain a NAC domain. Souer *et al.* (1996) described the first NAC gene as linked to the development of the shoot apical meristem and to the determination of the position of the meristems and primordia (first leaves after differentiation of the meristems) in petunia. NAC genes were then identified in all groups of terrestrial plants, halophytes but also in streptophyte green algae, which indicates that the emergence of NAC transcription factors even precedes the advent of land plants (Maugarny-Calès *et al.* 2016; Khedia *et al.* 2018).

Because of their role as transcription factors, NAC proteins regulate the expression of target genes through DNA-binding, either alone or by forming regulatory complexes with other proteins. A typical NAC protein contains, in its N-terminal region, a well-characterized, evolutionarily conserved DNA-binding domain of about 160 amino acids, called the NAC domain (InterPro; https://www.ebi.ac.uk/interpro/entry/InterPro/IPR003441/). It binds to the CACG core sequence in the promoters of certain target genes (Tran *et al.* 2004). In the C-terminal region, the NAC genes have a less studied, more diverged region of variable length, having a probable regulatory potential, which encompasses the activation, repression of transcription as well as protein binding (Olsen *et al.* 2005; Shao *et al.* 2015). This C-terminal region mostly comprises amino acid repeats and motifs enriched in serine-threonine, proline-glutamine or acidic residues (https://proteopedia.org/wiki/index.php/NAC_transcription_factor).

Nuruzzaman *et al.* (2015) showed that the expression of 75 NAC genes in rice is affected by at least one infection with one of the 5 viruses tested: Rice dwarf virus (RDV), Rice black-streaked dwarf virus (RBSDV), Rice grassy stunt virus (RGSV), Rice ragged stunt virus (RRSV), and Rice transitory yellowing virus (RTYV).

In wheat, certain NAC factors are also expressed in response to biotic or abiotic stresses. This is the case of *TaNAC8* and *TaNAC4*; the expression of these two genes is induced in the leaves in response to infection by *Puccinia striiformis* (agent of wheat yellow rust), but also to salt and cold stresses (Xia *et al.* 2010a, 2010b). TaNAC1 is also a negative regulator of resistance to fungal (*Puccinia striiformis*)[1] and bacterial (*Pseudomonas syringae*) diseases in wheat, acting upstream of the signaling pathways of jasmonic and salicylic acids (Wang *et al.* 2015).

1.3. MYB Transcription Factors

The MYB factors are chracaterized by the conserved MYB DNA binding domain usually containing one to four imperfect repeats of 50-53 amino acids, termed R1, R2, R3 and R4 (Tan *et al.* 2020). The first identified MYB gene was v-MYB, an oncogene originating from the Avian Myeloblastosis Virus [2]. MYB genes also exist in insects (Huang *et al.* 2012), fungi (Verma *et al.* 2017) and plants (Martin and Paz-Ares 1997). The first *MYB* gene cloned from a plant was the *C1* gene of maize (Paz-Ares *et al.* 1987); and plants possess a greater number of described *MYB* genes, compared to fungi and animals (Riechmann *et al.* 2000). For example, there are 168 estimated *MYB* genes *A. thaliana* and 489 in *Brassica napus* (Plant Transcription Factor Database v. 04; http://planttfdb_v4.cbi. pku.edu.cn/, accessed April 04, 2020).

In *A. thaliana*, *AtMYB44* is known to be involved in plant immunity to the green peach aphid, *Myzus persicae* (Liu *et al.* 2010). Likewise, *AtMYB102* has been reported to play a role in the the defensive response to the caterpillar, *Pieris rapae*

(De Vos *et al.* 2006). The role played by *MYB* genes in plant disease resistance was also revealed in other species. For example, recently, Zhang *et al.* (2019), using scanning electron microscopy and gas chromatograph–mass spectrometry analyses, found that a newly-identified MYB family member from apple (*Malus domestica*), *MdMYB30*, enhanced disease resistance[3] in this tree species, through the regulation of cuticular wax biosynthesis. Segarra *et al.* (2009) demonstrated that the MYB72 root-specific transcription factor occupies a central position in the signalling pathways prompted by two different (*i.e.* a fungal and a bacterial) symbiotic microorganisms, namely the fungus *Trichoderma asperellum* strain T34 and the rhizobacterium *Pseudomonas fluorescens* strain WCS417r. This demonstrates that MYB72 is able to activate the induced systemic resistance (ISR) triggered by these beneficial microbes.

1.4. AP2 / EREBP Transcription Factors

Transcription factors APETALA2 (AP2) and Ethylene-responsive element binding protein (EREBP) form a family of plant-specific transcription factors, the distinctive feature of which is the AP2 domain. AP2/EREBP genes are divided into two subfamilies: AP2 genes with two AP2 domains and EREBP genes with a single AP2/ERF (Ethylene Responsive Element Binding Factor) domain (Shigyo *et al.* 2006).

The AP2/EREBP genes are a large multigenic family involved in a number of physiological, biochemical and cellular adaptations, including the specification of organ and meristem identity during seed development, cell proliferation, secondary metabolism and responses to different plant hormones (Pande *et al.* 2018). Plants use these mechanisms to respond to various kinds of biotic and environmental stresses, such as drought, temperature and salinity (Dietz *et al.* 2010). AP2/EREBP transcription factor family is also responsive to cold stress with the help of C-repeat binding factors (CBFs), which bind to cis-elements in the promoter of cold responsive (COR) genes, enabling their transcription (Sakuma *et al.* 2002). It has also been shown that members of this family of transcription factors are involved in the jasmonic acid pathway and in the response to abiotic stress *via* the ABA biosynthetic pathway (Khong *et al.* 2008).

1.5. bZIP Transcription Factors

Basic leucine zipper (bZIP) family owes its name to the fact that the members contain a DNA-binding domain consisting of a 60–80 amino acids sequence, divided into two functional regions: an 18 aminoacids region (N-X7-R/K-X9) rich in basic amino acids, and a less conserved leucine-rich motif in which leucine occurs at regular intervals (leucine zipper) (Yang *et al.* 2019).

Findings in soybean indicated that four bZIP genes (*i.e.* GmbZIPE1, GmbZIPE2, GmbZIP105, and GmbZIP62) have a role in resistance to Asian soybean rust disease (ASR) (Alves *et al.* 2015). In cocoa (*Theobroma cacao*) two bZIP genes showed overexpressed following *Moniliophthora perniciosa* infection (Lopes *et al.* 2010). Likewise, nine bZIP family genes were upregulated by by *Ustilago maydis* infection (Wei *et al.* 2012a).

1.6. NPR1 Transcription Factors

NPR1, named after a mutant, *npr1* [also referred to noninducible immunity 1 (*nim1*) and salicylic acid-insensitive 1 (*sai1*)] is a component of the SA-dependent signaling pathway (Pieterse and Van Loon 2004). Analyzes of the *Arabidopsis npr1* mutant have shown that NPR1 is involved in the development of RH, and in basal resistance (Dong 2004). In the absence of biotic stress, the protein NPR1 is found in the cytoplasm in the form of oligomers linked by disulfide bridges. Yet, when plant cell is challenged by a pathogen, the accumulation of SA leads to a change in the redox state of the cell and to the reduction of cysteine residues, then to the monomerization of NPR1 (Mou *et al.* 2003). The monomers of NPR1 are then translocated in the nucleus, where they come into contact with a transcription factor TGA (belonging to the group of transcription factors bZIP) (Johnson *et al.* 2003). The transcription factor TGA then induces the transcription of defense-related genes, such as *PR1a[4]* (Kesarwani *et al.* 2007).

2. REGULATION OF TRANSCRIPTIONAL COMPLEXES

Although the transcriptional regulation of the signaling pathways leading to the activation of defensive mechanisms begins to be deciphered gradually, the nature of the *regulators* controlling the launching and the implementation of the hypersensitive cell death program still remains not very well-known.

2.1. Direct Regulation of Transcriptional Complexes by Transcription Factors

In eukaryotes, the expression of protein-encoding genes is subject to very finely crafted and controlled regulation. The main level of control is the transcriptional level, and it is known that transcription factors (TFs) have an important, positive or negative, regulatory role at this stage. They can bind to the cis-regulatory sequences located in the promoter of their target gene *via* their DNA-binding domain. A number of studies have shown that the main differences between the two types of defense (PTI/ETI) are not very qualitative but rather *quantitative and/or temporal* (Katagiri 2004). This means that several different pathogen-derived molecules are activating a typical regulatory circuit. The variable outcomes consistent with each form of immunity are likely caused by the operation of activators and repressors of defense gene expression (Eulgem 2005).

2.2. Regulation of Transcriptional Complexes by MAPK Cascades

Phosphorylable proteins, MAP kinases (Mitogen-Activated Protein kinase), are among the key regulators of signaling pathways. MAP kinase is the last kinase activated in a phosphorylation cascade involving three kinases. This activation is due to a phosphorylation governed by a MAPK kinase (MAPKK), which is in turn activated by a MAPKKK. This type of cascade is effective and allows the plant cell to respond in a few minutes, by converting the signals generated by different receptors. Once activated, MAP kinases can be translocated in the nucleus where *they phosphorylate transcription factors*. Convergent findings in various plant species indicated that MAP kinases and WRKY transcription factors are involved together in the regulation of pathogen resistance-related genes (Asai *et al.* 2002; Xu *et al.* 2006).

2.3. Regulation of Transcriptional Complexes by Ca^{2+} signaling

Changes in ion fluxes are a very early response to the plant's specific recognition of the pathogen. They consist of an entry of calcium ions (Ca^{2+}) and protons (H^+) and an efflux of potassium ions (K^+). In *Arabidopsis*, the study of the *hlm1* mutant (HR-like lesion mimic) highlighted the role of a calcium-dependent ion channel: CNGC4 (Cyclic Nucleotide Gated Channel), upregulated in response to infection (Balagué *et al.* 2003). The mutant *hlm1* is affected in the control of HR and in the resistance to a broad spectrum of avirulent bacteria. Similarly, the *dnd1* (defense, no death) mutant corresponds to a mutation in the protein CNGC2, very phylogenetically close to CNGC4, and exhibits increased resistance to avirulent pathogens (Clough *et al.* 2000). These data clearly show the importance of ion fluxes in the early stages of signaling leading to resistance. In addition, the flow of calcium ions is particularly important because it acts as a second messenger in the activation of defense mechanisms. A peak of calcium ions is, therefore, required for the initiation of the oxidative burst in parsley (*Petroselinum crispum*) (Jabs *et al.* 1997). In addition, experiments carried out on parsley protoplasts have shown the activation of a membrane channel allowing the influx of calcium into the cell in response to the application of Pep-13, an elicitor derived from *Phytophtora sojae* (Zimmermann *et al.* 1997).

CONCLUSION

Improving plant stress tolerance/resistance by controlling the transcription of TF genes is now a trendy research area, since most of these genes are inducible under specific stress and, when expressed, can command plant defense signalling cascades. During the last two decades, considerable advancement has been achieved in transcription factor investigation. Many of the developments concern the role played by these expression regulators in the response to abiotic stress and in plant growth, and there are relatively fewer studies with regard to biotic stress. In order to gain a deeper knowledge of their involvement in biotic stress, it is important to recognize the interactive functional partners of transcription factors, implicated in the regulation of target genes under specific biotic stress.

[1] A plant pathogen that causes stripe rust on wheat, but has other hosts as well (*Basidiomycota*).

[2] The Myb family was named after the viral Myb (v-myb avian myeloblastosis viral oncogene homolog). The viral Myb (v-Myb) causes aberrant activation of other oncogenes, leading to myeloblastosis (myeloid leukemia) in chickens. The animal Myb genes are proto-oncogenes.

[3]*i.e.* the resistance to the bacterial pathogen strain *Pseudomonas syringae* pv. tomato DC3000 (Pst DC3000), and to the fungal pathogen *Botryosphaeria dothidea* (Zhang *et al.* 2019).

[4] PR1a is a member of the Pathogenesis-related (PR) proteins family. PR1a protein was first identified in tobacco mosaic virus (TMV)-infected tobacco leaves (van Loon *et al.* 2006). In rice, OsPR1a is induced by blast fungus infection (Agrawal *et al.* 2001). Further documentation about Pathogenesis-related (PR) proteins and their role in plant immunity is provided in Chapter 18 of this book.

CHAPTER 17

Insights into the Role of Epigenetics in Controlling Disease Resistance in Plants

Abstract: Plants are masters of epigenetic regulation. All of the major epigenetic mechanisms known to occur in eukaryotes are used by plants, with the responsible pathways elaborated to a degree that is unsurpassed in other taxa. DNA methylation occurs in plant genomes, in patterns that reflect a balance between enzyme activities that install, maintain, or remove methylation. Histone-modifying enzymes influence epigenetic states in plants and these enzymes are encoded by comparatively large gene families, allowing for diversified as well as overlapping functions. RNA-mediated gene silencing is accomplished using multiple distinct pathways to combat viruses, orchestrate development, and help organize the genome. The interplay between DNA methylation, histone modification, and noncoding RNAs provides plants with a multilayered and robust epigenetic circuitry that has a tangible impact on the control of plant genes conferring resistance to different biotic and abiotic stresses, either directly or indirectly. Eventually, plants with the most suitable epigenome may be subject to selection.

Keywords: Epigenetic control, Epigenetic modifications, Epigenome, MicroRNAs (miRNAs), Non coding small (sRNA), Plant DNA methylation changes, Transgenerational epigenetically acquired resistance, Transposable elements (TEs).

INTRODUCTION

The genetic information encoded by DNA is packaged in a structure called chromatin. All of the changes undergone by chromatin (DNA methylation, post-translational changes in histones) induce changes in the expression of genes transmissible by cell division. The study of the regulatory mechanisms related to the expression of genes has shown that the latter in addition to being under the control of regulatory sequences (involving nucleotide sequences upstream: place of fixation of factors modifying the access to transcriptional machinery), are also dependent on another level of regulation. This additional level is called *epigenetics* (Silveira *et al.* 2013). However, epigenetics is a recent discipline, which has been in full development since the 2000s, and studies are still essential to estimate the extent of the epigenetic phenomena and their role in the evolution of species.

Particularly in plants, epigenetics would play an important role in the response of the organism, leading to the establishment of an advantageous phenotype which could be permanently fixed to result in an adaptation (Schlichting and Wund 2014; Rey *et al*. 2016). A recent study in *Arabidopsis* has shown the strong link between the DNA methylation profiles in more than 1,000 collections and their adaptation to their original climate (Kawakatsu *et al*. 2016). Another study also showed, this time on oak, that the SNPs found in loci that contained DNA methylation variants (SMP) showed a greater differentiation, which is in agreement with a role DNA methylation in the local adaptation of plants (Platt *et al*. 2015).

Current research aims to better understand *the share of epigenetic mechanisms within the framework of the theory of evolution*, a subject which is currently controversial (Danchin*et al*. 2011; Miska and Ferguson-Smith 2016). Another very important goal of understanding epigenetic mechanisms is to improve cultivated plants (Rodríguez López and Wilkinson 2015 ; Bilichak and Kovalchuk 2016). Indeed, the ability to fix the memory of stress through generations but also the identification of genes affected during the stress response represent relevant tools for breeders (Rodríguez López and Wilkinson 2015). Moreover, an epigenetic component of complex traits (epiQTL) has been reported in *Arabidopsis* by exploiting the famous epiRIL (Cortijo *et al*. 2014). Some authors suggest to couple epigenetic data with emerging biotechnologies (Transcription Activator-Like Effectors (TALEs) - and nuclease-defective Cas9 (dCas9) -based designed transcription factor systems) to modify the transcription profiles on specific loci in order to generate plants more tolerant of environmental constraints (Moradpour and Abdulah 2019).

1. DNA METHYLATION

DNA methylation reveals common characteristics among plants and animals (Zhang *et al*. 2008; Lister *et al*. 2008; Zemach *et al*. 2010). The presence of methylation in the gene bodies [1] in the two kingdoms indicates that this phenomenon is an ancestral feature common to eukaryotic genomes (Zemach *et al*. 2010). In addition, preferential methylation of exons, in comparison to introns[2], also seems to be an ancestral phenomenon common to animals and plants (Feng *et al*. 2010). Interestingly, methylation in gene bodies shows a correlation with gene expression; for example, in *Arabidopsis*, genes methylated in their gene bodies are expressed constitutively, while genes methylated in the promoter are expressed specifically in certain tissues (Zhang *et al*. 2006). Overall, this correlation between the level of gene-body methylation and the level of gene expression is parabolic, which means that the genes expressed at an intermediate level are the most methylated (Zhang *et al.* 2006). All these observations indicate a certain role of methylation in gene expression (initiation, elongation, and transcriptional termination or even alternative splicing).

1.1. Reduced DNA Methylation and Defense-Related Genes Priming

In tobacco, the *NtAlix1* gene, which is potentially involved in programmed cell death after attack by the tobacco mosaic virus (TMV), is specifically expressed in hypomethylated *NtMET1* mutants (Wada *et al.* 2004). Likewise, in *Arabidopsis*, DNA methylation has been shown to be an important element in suppressing the development of *Agrobacterium tumefaciens* (Bond and Baulcombe 2015). Also, the *A. thaliana* mutant *nrpe1*, showing a global hypomethylation of DNA, is more resistant to the biotrophic pathogen *Hyaloperonospora arabidopsidis* (*Hpa*) whereas two hyper-methylated mutants were more susceptible to this pathogen. On the other hand, *nrpe1* was more susceptible to the necrotrophic fungus *Plectosphaerella cucumerina*. Conversely, DNA methylation had opposite effects on *ros1* mutant, showing overall hypermethylation of DNA. The latter is more sensitive to *H. arabidopsidis* while it is more resistant to *P. cucumerina* (López Sánchez *et al.* 2016).

There is converging evidence from several recent studies, suggesting that *reduced DNA methylation increases the responsiveness of the plant immune system* (Espinas *et al.* 2016). This 'priming' of plant defense allows a faster induction of defense-related genes after pathogen attack, resulting in an increased resistance (Furci *et al.* 2019).

1.2. Plant Methylation Changes During Pathogen Infection

The methylation rate in a plant can be changed by infection with different pathogens. There are multiple examples of *dynamic changes in DNA methylation* during pathogen infection. For example, in *Arabidopsis*, infection with *P. syringae* DC3000 induces hypomethylation of the genome (Pavet *et al.* 2006; Hewezi *et al.* 2017). Genome hypomethylation has also been observed in soybean during infection with *Heterodera glycines*[3] (Rambani *et al.* 2015). Similarly, Dowen *et al.* (2012) described numerous stress-induced differentially methylated regions in the *DNA methylome* of plants exposed to the biotrophic pathogen *Pseudomonas syringae* pv.*tomato* DC3000 (*Pst*).

1.3. Transgenerational Epigenetically Acquired Resistance

Findings from several recent studies (Luna *et al.* 2012; Slaughter *et al.* 2012; Furci *et al.* 2019) suggested that plants can transmit *epigenetically* defense traits into their progeny, priming the immune system of their offspring against diseases.

Furci *et al.* (2019) studied a population of over 100 *A. thaliana* lines that had the same DNA sequences but differed in patterns of DNA methylation. The experiments identified four pericentromeric DNA regions that were less methylated in lines with enhanced resistance to downy mildew[4]. This form of resistance did not appear to affect plant growth or their resistance to other diseases or environmental stresses.

2. TRANSPOSABLE ELEMENTS

Transposable elements (TEs) are present in the genome of most eukaryotes with a clear correlation between the amount of TE present in the genome and its size. Basically, TEs make up 30 to 90% of plant genomes; around 30% in the small genome of *Arabidopsis* (140 Mbp), more than 80% in the average genome of maize (2,700 Mbp) and around 90% in the large genome of wheat (17,000 Mbp) (Vitte and Bennetzen 2006). The transposition capacity of transposable elements gives them a mutating power within the genome. As a result, the expression of these TEs is very controlled *via* a very effective repression or silencing mechanism. While the mechanisms for maintaining and strengthening the silencing of transposable elements are well known (Plasterk 2002; Robert and Bucheton 2004), the reasons why a transposable element is repressed *via* a heterochromatin state of the region in which it is found are little known.

If the transposition of TE is generally deleterious for genes and organisms, cases of positive effect of insertions of transposable elements on the expression of defense genes have been observed. For example, transposition of the *renovator* transposon inside the promoter region of the rice blast disease resistance gene, *Pit*, has been shown to cause the transcription of this gene, and confer resistance to the pathogen *Magnaporthe grisea* (Hayashi and Yoshida 2009).

Transposable elements can also be a regulatory element in the response of plants to biotic stress *via variation in the level of R genes methylation*. For example, the analysis of the promoter region of the *RMG1* gene [5] (*Resistance methylated gene 1*) of *A. thaliana*, which contains transposable elements, has shown that TEs serve as a platform for controlling the expression of this gene. This regulation is executed by an antagonistic game of methylation in the basal state and demethylation during the response to biotic stress (Yu *et al.* 2013). Another example concerns the triple mutant *rdd* demethylases (*ros1, dml2, dml3*) from *Arabidopsis*, which shows an increased susceptibility to *Fusarium oxysporum*. This susceptibility is consistent with the repression of the expression of more than 200 genes involved in the response to biotic stress, showing transposable elements in their sequences (Le *et al*. 2014).

3. ROLE OF NON-CODING RNAS IN EPIGENETIC CONTROL

Non-coding RNAs constitute another level of epigenetic control (Esteller 2011). However, not all non-coding RNAs induce epigenetic modifications and have a role in post-transcriptional regulation, as it is the case of micro-RNAs (Esteller 2011). Maintaining stress over several generations could, thus, make it possible to select individuals who present the most suitable repertoire of small RNAs and the most suitable epigenome (Borges and Martienssen 2015). In addition, the spread of mobile non-coding RNA to neighboring cells and throughout the plant has been demonstrated (Pyott and Molnar 2015). Recent studies have highlighted the role of small RNA molecules (sRNAs) as mobile signals of gene inhibition, among other things by influencing the methylation of genes. The target genes have diverse functions, suggesting that sRNAs are essential for plant development. Small mobile RNAs are able to move into meristematic tissues where they can alter DNA methylation and gene expression, but are also meiotically heritable (Lewsey *et al*. 2016). In addition, heterochromatic short-interfering RNAs (het-siRNAs) control important epigenetic mechanisms such as imprinting [6] and paramutation (Borges and Martienssen 2015).

In addition to sRNAs, long non-coding RNAs (lncRNA) play a role in the activation or repression of target genes (Holoch and Moazed 2015).

CONCLUSION

Genes, particularly resistance genes, can be turned on or turned off by several types of chemical changes that do not change the DNA sequence such as DNA methylations and histone modifications. All these adjustments form *epigenetic traits* grouped under the term of *epigenome*. These signals can lead to modulating the expression of genes, without affecting their sequence. The phenomenon may be temporary, but there are long-term epigenetic changes that persist when the signal, which induced them, disappears. In addition, there is another level in epigenetic modifications, as there is an array of regulatory elements for the epigenetic processes themselves, which are activated in stress responses. These are either small mobile molecules called micro RNA (miRNAs) or small interfering RNAs, which are involved in a wide array of processes (*e.g.* methylation removal, chromatin assembly, and repression of protein formation).

[1] Methylation of gene bodies is opposed to the methylation of promoters.

[2] The preferential methylation of the exons means these have been found to be more enriched in methylation in CG sites compared to introns, in both animals and plants.

[3] The soybean cyst nematode (SCN).

[4] Caused by the biotrophic downy mildew pathogen *Hyaloperonospora arabidopsidis*.

[5]*RMG1* encodes a NB-LRR disease resistance protein with a Toll/interleukin-1 receptor (TIR) domain at its N terminus. *RMG1* is expressed at high levels in response to flg22 (refer to chapters 6 and 11). Expression of this gene is controlled by DNA methylation in its promoter region.

[6] Genomic imprinting is an inheritance process independent of the classical Mendelian inheritance. It is an epigenetic process that involves DNA methylation and histone methylation without altering the genetic sequence. These epigenetic marks are established (imprinted) in the germline (sperm or egg cells) of the parents and are maintained through mitotic cell divisions in the somatic cells of an organism. Forms of genomic imprinting have been demonstrated in fungi, plants and animals.

CHAPTER 18

Plant Defense Gene Expression and Physiological Response

Abstract: The perception of the pathogen signal by the host specific receptors and its transduction by various signaling pathways culminate in the synthesis and synchronized accumulation of defensive molecules some of which play a structural role while others exercise a direct antimicrobial function. Biochemical mechanisms include, among others, the synthesis of peptides and antimicrobial proteins, hydrolytic enzymes as well as the production of phytoalexins and secondary metabolites with high antimicrobial potential. This chapter provides a synthesis of the remarkable progress made in recent years in terms of understanding the mechanisms applied in the molecular and physiological alterations that occur during the defense response of plants.

Keywords: Defensins, Hypersensible response (HR), Protease inhibitors, Phytoalexins, Pathogenesis-related proteins (PRs), Secondary metabolites.

INTRODUCTION

From their first contact with plants, phytopathogenic microorganisms are confronted with the barriers of constitutive defenses, which prevent the pathogen from penetrating the tissues of the host. When the microorganisms come to the apoplasm[1], the plant triggers more advanced defense mechanisms: perception and recognition of its aggressor, transduction of cellular signals. Finally, the outcome of the interaction between a plant and a pathogen, *i.e.* resistance or disease, will be dependent on the effectiveness of the arsenal of activated defenses. On the other hand, the speed with which the plant's response is expressed is, at this stage, a crucial criterion in the outcome of a plant-pathogen interaction since it will determine the implementation of resistance or the expression of the disease.

1. HYPERSENSIBLE RESPONSE (HR)

HR is one of the most spectacular and effective plant defense mechanisms. It is generally associated with specific host resistance but also with certain types of non-specific resistance. HR is a genetically programmed form of cell death that

Dhia Bouktila & Yosra Habachi

takes place very quickly and is limited to the area infected with the pathogen. Two roles are assigned to this defense response. On the one hand, the death of the plant cells surrounding the pathogen would considerably disturb the development of the latter. In fact, it is found in an environment devoid of nutrients, at an unfavorable pH and humidity, and containing toxic compounds released following the death of cells. On the other hand, the HR would play a role of *alarm signal*, informing the other parts of the plant of the attack. This signal generated during HR would spread to the whole plant, authorizing the establishment of *systemic defenses*.

2. ENZYMES AND ENZYME INHIBITORS

Protease inhibitors (PIs) are defensive proteins characterized by their ability to inactivate proteolytic enzymes of endogenous or exogenous origin (Ryan 1990). On the one hand, they protect the plant against uncontrolled endogenous proteolytic activity and, on the other hand, they control the exogenous proteases secreted by herbivorous insects or pathogens. They are subdivided into four categories according to the type of proteases with which they are associated:

a. **Serine Protease Inhibitors**, very widely distributed in plants;
b. **Cysteine Protease Inhibitors**, also abundant in plants;
c. **Metalloprotease Inhibitors**, which are relatively rare in plants;
d. **Acid Protease-specific Inhibitors**, rarely present in plants (Selitrennikoff 2001).

In collaboration with other molecules such as *phytoalexins*, PIs play an important role in the defensive arsenal of plants (Schimoler-O'Rourke *et al*. 2001). Many studies have focused on the effect of ingestion of PIs on insect survival. All results obtained under experimental conditions converge on the idea that the PIs ingested by insect larvae are *antimetabolic* and *antinutritive* molecules, which inhibit digestive proteases, thus depriving the pests of the amino acids essential for their growth and development (Zhu-Salzman and Zeng 2015). However, the selection pressure, exerted by the plant, often leads to the establishment in insects and pathogens of various adaptive mechanisms allowing them to overcome the toxic, repulsive or anti-palatable[2] effects of defense molecules (Kessler and Baldwin 2002). The relationships between plants and their aggressors are therefore constantly evolving, which seems to be responsible for biodiversity on earth.

3. DEFENSINS

Plant defensins are small, highly stable, basic, cysteine-rich[3] peptides that are a part of the plant innate immune system. They are termed plant defensins because they are structurally related to defensins found in other types of organism, including humans (Wong *et al.* 2007). The first members of the family of plant defensins were isolated from wheat and barley grains in 1990 (Colilla *et al.* 1990). A query of the UniProt database (www.uniprot.org/) currently reveals publications of 1,929 plant defensins available for review from TrEMBL section and 390 reviewed, manually annotated plant defensins from SwissProt section and that the *Arabidopsis* genome alone contains more than 330 defensin-like (DEFL) proteins [4].

Unlike the insect and mammalian defensins, which are mainly active against bacteria, plant defensins, with some exceptions, do not have antibacterial activity (Stotz *et al* 2009). These exceptions include *Cp-thionin II* from cowpea (Franco *et al.* 2006), *DmAMP1* from *Dahlia merckii*, *CtAMP1* from *Clitoria ternatea*, *ZmESR-6* from maize (Balandín *et al.* 2005), fabatin from broad been (Zhang and Lewis 1997), *SOD2* and *SOD7* from spinach (Segura *et al.* 1998), *MtDef4* and *MtDef5* from *Medicago truncatula* (Sathoff *et al.* 2019), all of them reported to exhibit antibacterial activity against a range of Gram-positive and Gram-negative bacterial pathogens. In contrast, most plant defensins have previously been shown to primarily inhibit the growth of fungal plant pathogens, such as *ZmDEF1* from *Zea mays*, *Psd1* from *Pisum sativum*, *NaD1* from *Nicotiana alata*, *TPP3* from *Solanum lycopersicum*, *BjD* from *Brassica juncea* (Indian mustard), and *MtDef4* and *MtDef5* from *Medicago truncatula*.

Most plant defensins were isolated from plant seeds; for example in radish, defensin proteins represents 0.5% of the total protein in seeds, and the amount of released proteins is sufficient to suppress the fungal growth in the soil (Terras *et al.* 1995). Yet, defensins have also been identified in other tissues from a variety of plants, including leaves, pods, tubers, fruit, roots, bark and floral organs (cited in Stotz *et al* 2009). These peptides have a biotechnological potential as they can be overexpressed in transgenic crops, often resulting in improved resistance to pathogen as was the case in tobacco, tomato, oilseed rape, rice and papaya (cited in Stotz *et al* 2009).

Antifungal plant defensins spectrum of organisms and mode of action: Plant defensins differ considerably in the spectrum of organisms inhibited and modes of action. Plant defensins interactions with fungal-specific components can be active as follows (cited in Sathoff *et al.* 2019):

- Disrupting and permeabilizing fungal plasma membranes, for example by interacting with bioactive fungal plasma membrane resident phospholipids, hence inducing fungal cell death,
- Disruption of Ca^{2+} gradient essential for polar growth of hyphal tips,
- Binding to specific sphingolipids present in the fungal cell wall or plasma membrane of their target fungi,
- Some defensins enter into fungal cells and bind to fungal intracellular targets,
- Induction of the production of reactive oxygen species (ROS),
- Inhibition of fungal cell division.

4. PHYTOALEXINS

The role of phytoalexin in disease resistance has been one of the major developments in physiological plant pathology in the past 50 years and this mechanism of resistance has stimulated many reasearches.

Phytoalexins (from the greek phytos, plant, and alekein, repelling) are phytochemicals of low molecular mass (plant secondary metabolites) synthesized *de novo* by a plant, generally in response to biotic stress (typically an attack by microorganisms), or sometimes to an abiotic stress (toxic metals, detergents, cold) and accumulated temporarily by certain plant tissues at the site of the fungal or bacterial infection, or at the site of the imposition of stress. They have inhibitory activity against bacteria, fungi, nematodes, insects and toxic effects for the animals and for the plant itself (Arruda *et al.* 2016).

Phytoalexins can be phenolic compounds (stilbenes, flavonoids), alkaloids, terpenoids, apigenidin, luteolinidin, apigeninidin or polyacetylenes (Arruda *et al.* 2016). There are many different chemical forms, depending on the plant species, displaying a wide range of antimicrobial activities (Bizuneh 2020). Phytoalexins are found in particular in the skin of the red grape in the form of ***resveratrol*** (Jeandet *et al.* 1995; Chang *et al.* 2011).

In general, the antifungal effect of phytoalexins manifests through the inhibition of the germination of fungal spores or mycelial growth (Hasegawa *et al.* 2014). Some phytoalexins also exert an antimicrobial effect (Chalal *et al.* 2014). The importance of phytoalexins in the defensive strategy of plants has been exploited in biotechnology, by transforming tobacco, tomato and alfalfa plants with the gene coding for ***stilbene synthase***, an enzyme involved in the synthesis of a vine phytoalexin, ***resveratrol***. Having acquired the capacity to synthesize this phytoalexin, the transformed tobacco plants expressed resistance to *Botrytis cinerea* (Hain *et al.* 1993), while the transformed tomato and alfalfa plants were more resistant to *Phytophthora infestans* (Thomzik *et al.* 1997) and *Phoma* *medicaginis*, respectively (Hipskind and Paiva 2000). Likewise, transgenic pea plants (*Pisum sativum* L.) that have lost their ability to produce **pisatin**[5] are much less resistant to fungal infection (Wu and VanEtten 2004). More recently, Zernova et al. (2014) described the transformation of soybean hairy roots with both the peanut resveratrol synthase 3 (AhRS3) gene, and the resveratrol-*O*-meth yltransferase (ROMT) gene from *Vitis vinifera*. Overexpression of these two genes resulted in the production of resveratrol and its methylated derivative **pterostilbene,** which is normally not synthesized by soybean plants, along with a lower necrosis of the transformed tissues (only 0 to 7%) in response to the soybean pathogen *Rhizoctonia solani* compared to the wild-type ones, which exhibited about 84% necrosis.

5. PATHOGENESIS-RELATED PROTEINS (PRS)

In addition to phytoalexins, which play a crucial role in plant innate immunity, the defensive arsenal of plants also includes many defense proteins, including the PR proteins, which have been studied in-depth (Sels *et al*. 2008). These stress proteins, discovered in 1970 in tobacco by two research groups (Gianinazzi *et al*. 1970; Van Loon and Van Kammen 1970), differ from all other known proteins by their molecular weight (between 10 to 30 kDa), their solubility, their great resistance to proteolytic digestion and their mainly extracellular localization (Kauffmann *et al*. 1999).

The major criterion for a protein to be included in the classification of PR proteins was originally linked to its ***induction*** by biotic stress and the increase in its activity specifically during plant-microbe interactions, even if this induction was not generalized to all pathogens (Van Loon 1999). That is why van Loon *et al*. (2006) introduced the term ***inducible defense-related proteins*** to define the originally proposed PR proteins.

The classification of PR-protein was updated from time to time. They were originally discovered from virus-infected tobacco plants, and based on serological characteristics and sequence data, PR-proteins were classified into five families: PR-1 to PR-5. PR-proteins generally have two subclasses, a basic subclass and an acidic subclass (Kitajima and Sato 1999). Later, different groups of proteins were identified as PR-proteins and included in this family. It now comprises 17 families of PR-proteins, PR-1–PR-17 (Sels *et al*. 2008 ; Ali *et al*. 2018).

For some of these 17 classes of PR proteins (Table **1**), the biological function remains a mystery. Most of the PR proteins identified seem to have direct (eg. **chitinases** and **glucanases**) or indirect (eg. **osmotins** and **permatins**) antimicrobial potential (Kombrink and Somssich 1997). In addition, it is not excluded that certain PR proteins act in synergy. This is the case, for example, with chitinases and glucanases that are produced simultaneously to allow more efficient degradation of the wall of pathogenic fungi (Theis and Stahl 2004). Other PR proteins such as ***oxalate oxidases*** (PR-15) catalyze the oxidation of oxalic acid, a powerful fungal toxin, to carbon dioxide (CO_2) and hydrogen peroxide (H_2O_2) (Hu *et al*. 2003). Other PR proteins include ribonucleases, ***defensins*** and ***thionines*** (Thomma *et al*. 1998), ***lipid transfer proteins*** (José-Estanyol *et al*. 2004) as well as certain ***protease inhibitors*** (Pulliam *et al*. 2001).

Table 1. Classification of PR-proteins (From Moosa *et al*. 2018)[6]

Family	Size (kDa)	Member	Properties
PR-1a	15	Tobacco (PR-1a)	Antifungal
PR-2	30	Tobacco (PR-2)	b-1,3-glucanases
PR-3	25-30	Tobacco P, Q	Chitinases (I, II, IV, V, VI, VII)
PR-4	15-20	Tobacco R	Chitinases (I, II)
PR-5	25	Tobacco S	Thaumatine-like proteins (TLPs)
PR-6	8	Tomato inhibitor I	Proteinase inhibitor
PR-7	75	Tomato P69	Endoproteinase
PR-8	28	Cucumber chitinase	Chitinase (III)
PR-9	35	Tobacco (lignin-forming peroxidase)	Peroxidase
PR-10	17	Parsley (PR-1)	Ribonuclease-like proteins (RLP)
PR-11	40	Tobacco class-V chitinase	Chitinase (I)
PR-12	5	Radish RsAFP3	Defensin
PR-13	5	Arabidopsis Thi2.1	Thionin
PR-14	9	Barley LTP4	Lipid-transfer protein (LTP)
PR-15	20	Barley OxOa (germin)	Oxalate oxidase (OXO)
PR-16	20	Barley OxOLP	Oxalate oxidase-like (OXO)
PR-17	27	Tobacco PRp27	Antiviral and antifungal

CONCLUSION

Plants produce a complexity of secondary metabolites, which act synergistically to combat plant pathogens and can be used by biotechnology for the isolation, synthesis and transfer of specific active molecules. Identification of phytoalexins, defensins or PR proteins in a particular species is a promising tool for engineering plants with efficient defense against biotic stresses.

[1] The apoplasm designates the extracellular continuum formed by the pectocellulosic walls and the empty spaces between plant cells.

[2] A palatable food is a one that has a pleasant taste.

[3] Stabilized by eight disulfide-linked cysteins

[4] Data are from the current study. Uniprot database was accessed on March 09, 2020.

[5] Pisatin is the major phytoalexin made by the pea plant *Pisum sativum.*

[6] The original bibliographical references of each PR-protein family are reviewed in Moosa *et al.* (2018)

CHAPTER 19

Contribution of Genomics to the Study of Resistance in Cultivated Plants

Abstract: Nowadays, agricultural genomics, or agrigenomics (the application of genomics in agriculture), continues to drive sustainable productivity and offer solutions to the mounting challenges of feeding the global population. Omic sciences (genomics, transcriptomics, proteomics, metabolomics) open today opportunities to create plants with high yields, independently of biotic and abiotic stresses. Plant genomic research, genome-wide computational tools and association analyses can assist in the identification of Resistance gene analogs (RGAs) from strategic plant species, and the detection of disease resistance QTLs. This chapter summarizes some of the large-scale genomic tools and studies that have clarified the plant – pathogen interactions.

Keywords: Plant Genomic Research, Computational Analyses, Genome-Wide Analyses, Big Biological Data, Agrigenomics, Biotic Stresses, R-gene Analogs (RGAs), NBS-LRR-encoding Genes, TILLING, RNA-seq, GWAS.

INTRODUCTION

The origin of the term genomics is recent since it was proposed by Tom Roderick in 1989 to designate the science having for subject the study of genomes. This new discipline aims to identify all the genes of a living organism. The first sequence of a genome, that of the bacteriophage PhiX174 (5386 bp) was published in 1977 (Sanger *et al.* 1977). Several virus, chloroplast and mitochondrial genomes were then sequenced for the next 20 years.

The advent and improvement of next-generation sequencing (NGS) technology has rapidly expanded the genomic information of numerous organisms and accelerated the generation of ***multiomic*** (genomic, transcriptomic, proteomic and metabolomic) data, leading to a new era of '***big biological data***'.

For plants, in particular, genetic resources are crucial for crop-breeding programs. Rich plant genetic resources have become available. As of March 2020, the genomes of ~ 100 angiosperm species have been completely sequenced and their genome data are hosted in well-constructed customized databases. Most of these species are plants of high economic importance or their wild relatives. Never-

Dhia Bouktila & Yosra Habachi

theless, the number of sequenced plant species (without customized databases) is much higher and is above 230 angiosperms (Chen *et al.* 2018).

A crucial challenge that emerges from genome data availability, resides in structuring (integrating and organizing) these plant omic data and linking them to particular phenotypes, which will contribute significantly to broadening and deepening our understanding of the molecular and genetic mechanisms that underly plant growth and adaptation to surrounding constraints, including abiotic and biotic stresses, therefore efficiently assisting in the breeding programs of major agricultural crops.

1. PLANT GENOMIC RESEARCH

In agriculture, the sequencing of the genomes of the main economic food crops and livestock is a project that has been going on for a long time. In particular, plant genomics was kick-started in 2000, when the common weed, *Arabidopsis thaliana*, was sequenced and promoted to a status of celebrity, as a model species. However, advances in many important plant species were hindered by the complexity of their genomes. It took almost 20 years for most of the genomes of agricultural crops around the world to be sequenced. Almost four years after the launch of an international rice genome sequencing consortium (International Rice Genome Sequencing Project, IRGSP), the rice genome was completely sequenced in 2002 (Goff *et al.* 2002). It is the second plant genome, after that of *A. thaliana* published in 2000. In 2005, an international consortium for the sequencing of the wheat genome (International Wheat Genome Sequencing Consortium, IWGSC) was launched. A draft of this genome was published in November 2012 (Brenchley *et al.* 2012). This research was particularly long to carry out because of the complexity of the wheat genome, comprising between 94,000 and 96,000 genes, five times more than that of humans. Just one year later, Chinese teams published the genomes of two species of spontaneous wheat; *Triticum urartu* (the supposed donor of genome A to cultivated wheat hexaploid) (Ling *et al.* 2013) and *Aegilops tauschii* (the supposed donor of genome D to cultivated wheat hexaploid) (Jia *et al.* 2013). At the time of writing these lines, the latest publication reporting a whole-genome sequence of a plant species is Chen *et al.* (2020) (Epub ahead of print, 20 april 2020) reporting the genome sequences of five cotton (*Gossypium*) allopolyploid species.

Since the genome sequence of the model plant *Arabidopsis thaliana* was published in 2000 (Arabidopsis Genome Initiative 2000), around 100 plant genome sequences have been published (plaBi database; https://www.plabipd.de /portal/web/guest/home1)[1]. To this number, we can add that of plant transcriptome assemblies available. As of March 2020, the One Thousand Plant Transcriptomes

Initiative database (One Thousand Plant Transcriptomes Initiative 2019) (https://sites.google.com/a/ualberta.ca/onekp/) includes >1300 plant transcriptomes[2].

Genomic analysis of an organism makes it possible to understand its cellular physiology, to grasp a very large number of its biological processes and metabolic activities, either by experimental evidence, or by *analogy with a model system*. The sequencing of complete genomes can also highlight *gene transfers* (Quispe-Huamanquispe *et al.* 2017) and allow a better understanding of the coevolutionary mechanisms of virulence and immunity.

Agrigenomics will help us develop new varieties of food crops with better nutritional content and *better resistance against pathogens*. However, genomics has also a variety of vocations. For example, genomics will help improve the environment by defining how diseases work in forest crops and by using the traits of living organisms to clean up polluted environments. Industrial biotechnology will continue to use genomics (and proteomics) to create many useful products, such as biofuels, antibodies, vaccines, plastics and biodegradable cosmetics.

2. FROM PLANT GENOMES TO PLANT PHENOTYPES: THE ANNOTATION OF PLANT GENOMES AS A FIRST STEP INTO THE IMPROVEMENT OF PLANTS FOR RESISTANCE TO BIOTIC STRESSES

Next-generation sequencing has triggered an explosion of available genomic and transcriptomic resources in plant sciences. However, whole genome sequencing is still only the first step to identifying the function of genes present in the genome of a plant species. Other tools are necessary for the precise identification of the function of these genes, including the functional annotation of genes.

Especially regarding plant disease resistance genes, pinpointing the mechanism and causes of disease requires more effort than sequencing a specific genome. For that, bioinformatics and computing technology is helping to unravel the coding DNA and predict gene sites in the genome as well as gene function based on similarity to other deposited sequences, which should provide many insights. However, similarity does not always translate into equivalent function especially when comparing across distant taxa. Furthermore, large portions of sequenced genomes in plants have not yet been assigned a putative function based on homology to known proteins.

3. GENOME-WIDE IDENTIFICATION OF *R* GENE ANALOGS (RGAS) FROM PLANT SPECIES

Despite the considerable amount of genomic resources available, a comparatively limited number of *R*-genes have been cloned and studied in detail, generating information on their structure, function and evolution, and providing useful genetic resources to develop new resistant cultivars. In contrast, thousands of RGAs have been identified in many plant genomes. Many computational biology-based genome-wide investigations of NBS-LRR family in various plants species have been published, which includes *Arabidopsis thaliana* (Meyers *et al.* 2003), *Solanum tuberosum* (Lozano *et al.* 2012), *Vitis vinifera* (Yang *et al.* 2008), *Brachypodium distachyon* (Tan and Wu 2012), papaya (Porter *et al.* 2009), kiwi (Li *et al.* 2016), *Rosaceae* (Arya *et al.* 2014), *Populus trichocarpa* (Kohler *et al.* 2008), *Medicago truncatula* (Song and Nan 2014), Pineapple (Zhang *et al.* 2016), *Cucumis sativum* (Wan *et al.* 2013), Soybean (Kang *et al.* 2012) and olive (Bettaieb and Bouktila 2020). All these studies have led to the identification of hundreds of NBS-LRR encoding genes from these plant genomes (Table 1) and have shown that these NBS-LRRs are highly duplicated and evolutionarily diverse. Angiosperms possess NBS-LRR-encoding genes but TNL encoding genes are known to be absent from grasses (McDowell and Simon 2006) and other monocots (Tarr and Alexander 2009; Xue *et al.* 2020) genomes.

Table 1. Genome-wide identification of RGAs in plant genomes (from Sekhwal *et al.* 2015).

Species	Genome Size (Mb)	Total Annotated Genes	NBS Coding Genes [a]							RLK [b]	RLP [c]	Other [d]
			CNL	TNL	CN	NL	TN	N	Total			
Dicots												
Arabidopsis thaliana (Arabidopsis)	125	25,498	51	79	8	20	17	26	201	600	56	46
Arabidopsis lyrata (lyrata)	207	32,670	21	103	17	14	20	10	185	-	-	-
Populus trichocarpa (black cottonwood)	485	45,555	119	64	19	83	13	46	344	379	-	127
Vitis vinifera (grape)	475	30,434	203	97	26	12	14	0	352	-	-	210
Linum usitatissimum (flax)	373	43,484	31	57	10	5	22	7	132	-	-	16
Solanum lycopersicum (tomato)	900	34,727	118	18	19	43	5	49	252	16	13	13

(Table 1) cont.....

Species	Genome Size (Mb)	Total Annotated Genes	NBS Coding Genes [a]							RLK [b]	RLP [c]	Other [d]
Carica papaya (papaya)	372	28,629	4	6	-	-	-	44	54	-	-	-
Cucumis sativus (cucumber)	367	26,682	25	19	1	17	5	3	70	-	-	-
Solanum tuberosum (potato)	844	39,031	65	37	24	184	12	113	435	-	-	142
Medicago truncatula (Medicago)	454	62,388	152	118	25	0	38	328	661	-	-	92
Gossypium raimondii (cotton)	880	40,976	35	41	18	96	9	31	230	60	144	56
Brassica rapa (chinese cabbage)	485	41,174	19	93	15	27	23	29	206	-	-	42
Brassica oleracea (cabbage)	630	45,758	6	40	5	24	29	53	157	-	-	82
Fragaria vesca (strawberry)	240	34,809	-	61	-	16	8	1	86	-	-	8
Malus x domestica (apple)	742	57,386	218	161	54	276	69	182	960	-	-	110
Lotus japonicus (lotus)	472	19,848	9	8	19	3	16	29	84	-	-	-
Theobroma cacao (cocoa)	430	28,798	82	8	46	104	4	53	297	-	-	17
Physcomitrella patens (moss)	510	35,938	9	3	2	5	0	1	20	-	-	45
Average	500	37,433	69	56	19	55	18	56	263	264	71	72
Monocots												
Oryza sativa (rice)	420	59,855	159	0	7	40	3	45	254	1429	90	281
Triticum aestivum (wheat)	17,000	94,000	98	-	0	555	-	318	971	-	-	1266
Zea mayes (maize)	2300	32,540	58	0	21	31	0	69	179	113	-	2
Sorghum bicolor (sorghum)	739	34,496	36	0	99	133	0	64	332	-	-	114
Hordeum vulgare (barley)	5100	30,400	101	-	51	145	-	34	331	-	-	89
Brachypodium distachyon (Brachypodium)	272	25,532	133	0	28	87	0	34	282	-	-	34

(Table 1) cont.....

Species	Genome Size (Mb)	Total Annotated Genes	NBS Coding Genes [a]							RLK [b]	RLP [c]	Other [d]
Triticum urartu (Red wild einkorn)	4940	34,879	235	0	44	218	-	38	535	-	-	35
Aegilops tauschii (Tausch's goatgrass)	4360	43,150	296	0	63	288	-	81	728	-	-	112
Average	4391	44,357	140	0	39	187	1	85	452	771	90	242

a : CNL, CC-NBS-LRR; TNL, TIR-NBS-LRR; CN, CC-NBS; NL, NBS-LRR; TN, TIR-NBS; N, NBS; b: RLK, receptor like kinase; c: RLP, receptors like proteins; d: Other, includes TIRX, XN, TNLX, TNTNL, TTNL, XTNX, CNX, TX and Partial NBS–LRR.

Plant RGAs are potential *R*-genes that have conserved domains and structural features which have specific roles in host-pathogen interactions. Mining and characterizing genome-wide plant RGAs using computational approaches are rendered possible due to their significant structural features and conserved domains. Enormous advances have been made in our knowledge of *R*-genes and in elucidating the role and ***mechanism of action*** of genes involved in plant defense response[3].

4. TARGETING INDUCED LOCAL LESIONS IN GENOMES (TILLING)

This is a new technique (McCallum *et al.* 2000) that could pave the way for a new green revolution. By mixing random chemical mutagenesis and PCR-based screening, TILLING (Targeting Induced Local Lesions in Genomes) improves a character in a plant without creating transgenic plants. The principle consists in mutating thousands of plants at random and screening them in order to identify, at the molecular level, the one with the desired mutation. Instead of carrying out the gene sequencing in several plants of each family, which would be heavy and expensive, the DNA of each family is hybridized to an unmutated control DNA. If there is a difference in sequence between the control DNA and the DNA tested, the hybrid DNA obtained will have a mismatch which will be detected by a very sensitive biochemical method. This technique makes it quite easy to obtain a collection of mutations affecting a particular gene. The plants identified as mutants are then studied for their phenotype (Fig. **1**).

Since the advent of TILLING, this method has been widely used for the study of functional genomics in plants, especially for the model plant *A. thaliana,* with the *Arabidopsis* TILLING Project (ATP) (Till *et al.* 2003). Another model plant, *Lotus japonicus*, has also been the focus of elucidating gene function through TILLING. For example, nitrogen fixation and the functional role of sucrose synthase were the target of a *Lotus japonicus* TILLING study performed by Horst *et al*. (2007). In pea (*Pisum sativum*), another nitrogen-fixing plant and a member of the legume family, TILLING was applied to identify an allelic series of mutations in five genes with a total of 60 mutants identified (Triques *et al*. 2007). Barley, which is also an important cereal crop with a fairly large genome size of ~5,300 Mb, was evaluated for the ability of induced mutations to be detected by TILLING (Caldwell *et al*. 2004).

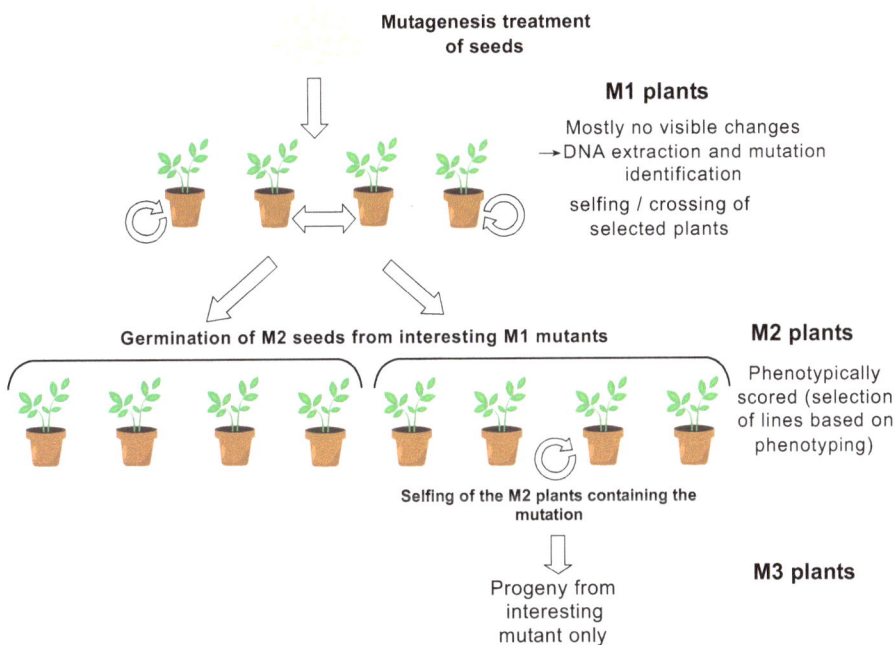

Fig. (1). The population design of TILLING. Seeds are mutagenized using a variety of chemicals and sown, resulting in M1 plants. All M1 plants are analyzed for DNA in order to determine interesting mutations (*e.g.* non-synonymous mutations) resulting in phenotypic alterations in the plant. The M1 plants are open- or self-pollinated in order to produce the M2 plants containing the mutation.

TILLING offers at least two advantages: (a): Europe refuses transgenic plants (Key *et al*. 2008). This method thus makes it possible to improve a character without going through genetic engineering; besides (b): the efficiency of mutagenesis is independent of the size of the genome studied (Stemple 2004; Kurowska *et al*. 2011).

5. TRANSCRIPTOME ANALYSIS

The transcriptome is the set of RNAs derived from the expression of part of the genes of the genome, in a cell or tissue type, at a specific time, and under given conditions. The characterization and quantification of the transcriptome in a biological model makes it possible to identify the genes transcribed in a given context and thus to determine the mechanisms of regulation of gene expression (or co-expression) and to define their regulatory networks. Better knowledge of the level of expression of a gene in different situations is a step forward towards understanding its function, but also towards the screening of new molecules and the identification of new diagnostic tools.

Introduced in the 1980s, the DNA microarray technology have made it possible to simultaneously measure the level of expression of a large set of messenger RNAs contained in a sample, making it a tool of choice for the study of the transcriptome. This method is still widely used today to understand multiple biological processes in plants, such as circadian clock, *plant defense*, environmental stress responses, fruit ripening, seed development, and nitrate assimilation (Aharoni and Vorst 2002).

Currently, with the development of very high throughput sequencing, new techniques for studying the transcriptome have emerged; these are, in particular, the *RNA-seq* (Wang *et al.* 2009) and the SAGE-seq (Velculescu *et al.* 1995). The RNA-seq (*Whole Transcriptome Shotgun Sequencing*) is a transcriptomic tool allowing the sequencing of all the transcripts of a sample. The SAGE-seq (*Serial Analysis of Gene Expression*) also called DGE-seq for *Digital Gene Expression* is a technique allowing the analysis of the level of expression of a large number of genes *via* the identification of *SAGE tags* derived from untranslated regions (5'UTRs or 3'UTRs) and their counting. SAGE-Seq is a very effective tool for in-depth transcriptome analysis, which has been developed to quantitatively assess the expression patterns of thousands of genes without prior sequence knowledge.

Similar to DNA microarrays, transcriptome analysis by very high throughput sequencing has quickly become a valuable asset for studying the resistance of plants to disease and pests. In this context, we can cite the following works using these high-throughput transcriptomic approaches:

a. RNA-Seq has been successfully used in many plants such as *Brachypodium sylvaticum* (Fox *et al.* 2013), *Sorghum sudanense* (Li *et al.* 2016), sugarcane (Cardoso-Silva *et al.* 2014), pepper (Ashrafi *et al.* 2012), orchardgrass (Huang *et al.* 2015), *Hemarthria* (Huang *et al.* 2016), and annual ryegrass (Pan *et al.* 2016).

b. Zhou *et al.* (2015), using RNA-seq, produced the first transcriptome of the tropical plant, *Bombax ceiba*, and listed 59 differentially expressed genes with greater than 1,000-fold changes under two conditions. Key genes with high RPKM[4] gene expression levels in drought, *AUX1*, *JAZ*, and *psbS*, are implicated in the control of plant growth, resistance against abiotic stress, and the photosynthesis process.

c. Venu *et al.* (2007) detected numerous up- and down-regulated rice genes after infection with *Rhizoctonia solani*[5] using SAGE and microarray analysis.

6. CONTRIBUTION OF GENOME-WIDE ASSOCIATION STUDIES (GWAS) TO THE IMPROVEMENT OF PLANTS FOR RESISTANCE TO BIOTIC STRESSES

Genome-wide association studies aim to determine links between a disease and some genetic markers. Even though plant genomics and genome-wide sequences are becoming more and more accessible, the use of this genomic information, to create selective advantages for improvement, is still far below the technology allowing the production of sequence data.

Genetic markers are tiny DNA segments scattered across the genome. They are successfully deployed to genotype individuals and to identify those that contain agronomic traits of importance to be exploited in breeding programs. Therefore, genetic markers help accelerate selection and genetic gain.

Genome-wide association studies (GWAS) scan the entire genome in order to identify markers explaining part of the phenotypic variability observed for traits of interest (Li *et al.* 2019). The multi-locus mixed-model (MLMM) method (Segura *et al.* 2012) uses a stepwise forward-backward type algorithm, allowing the integration of relevant associations in the association model. This approach increases the power of the association study, and provides the set of markers that best explain the observed variability (Segura *et al.* 2012).

Several *recent studies* by GWAS have been carried out in several species, which provided valuable information for resistance to pathogens:

a. Potnis *et al.* (2019) conducted a comprehensive study by GWAS on a varietal collection of ***pepper (Capsicum)***, to identify new sources of resistance to *Xanthomonas gardneri*. The QTLs resistant to *X. gardneri*, identified in this study, provided alleles, which could be used for pyramiding resistance genes against different species of *Xanthomonas* in pepper.

b. In ***rice***, Li *et al.* (2019) conducted a genome-wide association study, which identified 56 QTLs of resistance to the fungus *Magnaporthe oryzae*. One of the QTLs is localized with the resistance gene *Pik* locus conferring resistance to *M. oryzae*. This new QTL corresponded to a new allele, named *Pikx*, at the *Pik* locus.

c. GWAS has also been used by Desgroux *et al.* (2016). The genotyping of 175 lines of ***Pisum sativum***, identified 52 QTLs associated with resistance to *Aphanomyces euteiches* and validated six of the seven QTLs previously reported.

d. GWAS has also been applied to detect SNPs significantly associated with resistance to *Heterodera glycines* in a ***bean*** collection. 84,416 SNPs were identified in 363 bean accessions (Wen *et al.* 2019).

CONCLUSION

The genomic era, in which high-throughput and cost-effective genomic tools are available, is revolutionizing our understanding of the complex interactions between plants and pathogens. Genomic approaches include TILLING, which enables screening mutagenized germplasm collections for allelic variants in target genes. Genome-wide association studies are very useful for the genome-wide discovery of markers. All these tools and advances in genomics are providing breeders with new tools and methodologies that allow a great leap forward in plant breeding for disease resistance.

[1] Phytozome 12 (https://phytozome.jgi.doe.gov/) and EnsemblPlants (http://plants.ensembl .org/) contain 93 and 71 sequenced genomes, respectively. All databases (PlaBi, Phytozome and EnsemblPlants) were accessed in March 2020.

[2] 1,341 entries from 1,124 species that span the diversity of plants. Information accessed in March 2020 (http://www.onekp.com/samples/list.php).

[3] Refer to Chapter 14 for more detailed documentation.

[4] RPKM, which stands for **Reads Per Kilobase Million**, is a metric of gene expression in RNA-Seq, normalized for sequencing depth and gene length.

[5] The fungal pathogen *Rhizoctonia solani* causes sheath blight, a major disease in rice. It also causes devastating diseases in hundreds of other host plants, including soybean, bean, sorghum, corn, and sugarcane.

CHAPTER 20

State of the Art and Perspectives of Genetic Engineering of Plant Resistance to Diseases

Abstract: The improvement of plants, in order to give them better resistance to diseases, relies on the existence of diverse natural populations. Traditionally, the breeder's role has been to crossbreed populations to obtain varieties possessing the desired traits. Modern advances in genetic engineering, in association with omic sciences, allow breeders to increase the genetic diversity of the populations on which selection operates, and to introgress new traits using various molecular methods, such as chemical or irradiation-based mutagenesis, genetic transformation or genome editing. It is most often a question of modifying existing varieties in order to obtain new ones which have the properties desired by researchers, according to the needs expressed by the various actors concerned. Here, we review the increasing usefulness and applicability of biotechnology and genetic engineering approaches for accelerating variety development and crop improvement.

Keywords: Bacterial Harpins (*hrp*) genes, *Bacillus thuringiensis* (*Bt*) delta-endotoxin genes, Biotechnology, Crop improvement, CRISPR-Cas9, Genetic Engineering, Genetically engineered crops, Plant Protease Inhibitors (PI), Plant genome editing, RNA-based antiviral resistance.

INTRODUCTION

Conventional breeding (based on intervarietal crosses) plays an essential role in crop improvement but generally involves the examination of large populations of crops over several generations, which is a long and laborious process. Genetic engineering, which refers to the direct modification of an organism's genetic material, using biotechnology, offers several advantages over conventional breeding. First, crops with the desired agronomic characteristics can be obtained in fewer generations than conventional breeding. Second, the genetic material that can be exploited for the introgression, deletion, modification or regulation of genes of specific interest is not limited to the genes available within the target species, because genetic engineering allows the transfer of genetic material between distinct species, even between distant or belonging to different kingdoms. Finally, genetic engineering makes it possible to improve vegetatively propagated plants, such as the banana tree (*Musa* sp.) and potato (*Solanum tuberosum*). These characteristics make genetic engineering a powerful tool for improving resistance to plant pathogens.

In addition, the induction of mutations in plants is made possible by two main methods, namely irradiation and treatment with chemical mutagens (Leitao 2011). The most commonly used chemical mutagens cause, almost strictly, single base substitutions. The main benefit of the chemical mutagens is that they can be used to prepare mutant populations with high mutation levels, which facilitates the detection of specific mutations in a population (Szarejko *et al.* 2017). A major change in the use of mutants occurred following development of effective TILLING [1] (Targeting Induced Local Lesions in Genomes) techniques (McCallum *et al.* 2000). Actually, TILLING made reverse genetics approaches pertinent, because this technique is intended to detect mutations in specific, known genes. This now makes it easier to find mutations in any pre-defined gene from across the genome of a crop plant, if the DNA sequence of the gene is identified and if an appropriate mutant population is accessible (Jankowicz-Cieslak *et al.* 2017).

All these techniques set a new scene for assessing the precision of new breeding techniques.

1. GENETICALLY ENGINEERED CROPS

The first transgenic plant was produced in 1983, leading to a considerable ***paradigm shift*** [2]: the desired agronomic trait may no longer be selected but ***introduced*** by molecular techniques. The individual produced by transgenesis is called a genetically modified plant (GMP). GMPs are produced by the introduction of DNA fragments carrying the sequence corresponding to the gene encoding the desired trait. The gene can come from the plant, animal or microorganism kingdoms. It can even, depending on the case, be of synthetic origin.

1.1. Broad-Spectrum Resistance Conferred By PRR and Chimeric PRR Transgenes

It has been suggested that the PRR-mediated recognition of PAMPs can be used, in resistance engineering, to confer a ***wider spectrum of disease resistance*** to transgenic plants (Borras-Hidalgo *et al.* 2012). For instance, the *Arabidopsis* EFR, which recognizes the bacterial elongation factor EF-Tu, has been shown to confer resistance against a number of bacteria when transferred into Solanaceae species

(Lacombe *et al.* 2010). It has also been demonstrated that chimeric PRRs may be used to develop a double resistance to both bacteria and fungi (Brutus *et al.* 2010). The combination of different PRRs and chimeric PRRs in a single plant, which recognize several non-self structures, is likely to be the best way to build broad-spectrum and more durable disease resistances.

1.2. Plant Defensins Transformed into Target Plants

Plant defensins have a biotechnological potential because they can be overexpressed in genetically egineered crops, often resulting in enhanced resistance to pathogens as was the case in tobacco, tomato, oilseed rape, rice and papaya (cited in Stotz *et al.* 2009).

Jha and Chattoo (2010) transformed the peptide Rs-AFP2[3] into rice (*Oryza sativa* L. cv. Pusa Basmati). The transgenic plants were tested *in vivo* and *in vitro* against *Magnaporthe oryzae* and *Rhizoctonia solani*, the main causes of rice losses in agriculture, revealing that overexpression of Rs-AFP2 can control the rice blast and sheath blight diseases (Jha and Chattoo 2010). Another example is the transformation of tobacco with the mustard defensin, BjD, which once more validated the potential of these peptide-family members as excellent antifungal agents, as transgenic plants displayed improved resistance towards *F. moniliforme* and *Phytophtora parasitica* (Anuradha *et al.* 2008). More recently, a defensin purified from maize, ZmDEF1, when transformed into tobacco plants, showed increased tolerance against *Phytophtora parasitica* (Wang *et al.* 2011).

1.3. Plant Protease Inhibitors (PI) Transformed into Target Plants

Plant protease inhibitors (PI) are able to protect plants against insect attacks by interfering with the proteolytic activity of insects' digestive gut. This mechanism contributes to defenses against many herbivorous arthropods and microbial pests (Rustgi *et al.* 2017). In particular, serine and cysteine PIs are abundant in plant seeds and storage tissues (Reeck *et al.* 1997) and may contribute to their natural defense system against insect predation. The first PI gene that was successfully transferred artificially to plant species resulting in enhanced insect resistance was isolated from cowpea and encoded the trypsin/trypsin inhibitor CpTI (Cowpea Trypsin Inhibitor) (Hilder *et al.* 1987). Oryzacystatin 1 (OC1) is a well-studied cysteine PI from rice seeds which has been successfully introduced into several different crops like rice (Duan *et al.* 1996), wheat (Altpeter *et al.* 1999), oilseed rape (Rahbe *et al.* 2003) and eggplant (Ribeiro *et al.* 2006). It protects these plant species against beetle attacks and, in some cases, aphids (Sharma *et al.* 2004).

1.4. Bacterial Harpins (*hrp*) Genes Transformed into Target Plants

The Harpins (*hrp*) genes encode type III secretory pathways and many phytopathogenic bacteria require them either for eliciting hypersensitive responses (HR) on resistant host plants or for pathogenesis on susceptible hosts. When Harpins (*hrp*) proteins are secreted by pathogenic bacteria into the plant cells, localized cell death happens through a series of reactions such as reactive oxygen species (ROS) accumulation. This strategy was exploited to produce transgenic plants resistant to bacterial pathogens. Harpin NEa (HrpNEa) is encoded by the gene *hrpN* located on the chromosome of *Erwinia amylovora* causing the fire blight disease of apple[4] (Kim and Beer 2000). HrpNEa is a documented inducer of systemic acquired resistance (SAR) in plants. It has been shown that increased HrpNEa levels trigger boosted resistance to fire blight in transgenic pears, *Pyrus communis* (Malnoy *et al.* 2005).

1.5. *Bacillus thuringiensis* Delta-Endotoxin Genes Transformed into Target Plants

The know-how of genetic engineering has recently opened up new possibilities for intervention through genetic transformation of target plants by insertion and expression in their genome of genes encoding insecticidal toxins. The example is given by plants genetically transformed by insertion and expression of the ***delta-endotoxin gene*** from *Bacillus thuringiensis* (*Bt*). This soil bacterium has the particularity of synthesizing entomopathogenic toxins during sporulation, particularly a delta-endotoxin degrading the intestinal epithelium cells of caterpillars or mosquito larvae. After ingestion in the intestinal epithelium, it causes almost immediate paralysis of the digestive tract and mouthparts. This paralysis obviously leads to stopping feeding, then the death of the insect. This delta-endotoxin is completely harmless to humans and other vertebrates.

From 1996 to 2013, farmers worldwide have planted a total of 570 million hectares of genetically engineered crops producing insecticidal proteins from the bacterium *Bacillus thuringiensis* (*Bt*) (Tabashnik and Carrière 2015). Currently, *Bt* crops are reported to occupy more than 43% of global biotech areas (James 2015). There are also several variants of toxins naturally produced by *Bt* bacteria that harm different groups of insects. For example, the toxin ***Cry1Ab***, one of the most widely used toxins in genetic engineering, harms Lepidoptera, but not insects of other orders. There are a multitude of plant species genetically modified to contain the *Bt* toxin, but the most common *Bt* crops remain corn and cotton (Abbas 2018):

- ***Bt* potato** (Rahnama *et al*. 2017),
- ***Bt* corn**: Cultivation of *Bt* corn started in the USA, Canada, and Europe (Spain) in 1997, and by 2009, it was commercially planted in 11 countries. It was then representing 85% of the total area of corn in USA, 84% in Canada, 83% in Argentina, 57% in South Africa, 36% in Brazil, 20% in Spain, and 19% in Philippines (cited in Abbas 2018). In 2016, GM corn in the world (in 16 countries) reached 60.6 million ha, out of which 6 million (10%) were Bt corn (cited in Abbas 2018).
- ***Bt* cotton** (Qiao 2015),
- ***Bt* tomato**, Transgenic tomato line expressing modified *Bacillus thuringiensis* *Cry1Ab* (Koul *et al*. 2014) and *Cry2Ab* (Saker *et al*. 2011) genes have been produced and showed resistance to lepidopteran pests.
- ***Bt* soybean** (Martins-Salles *et al*. 2017),

Studies carried out in France (Bourguet *et al*. 2002), the United States (Obrycki *et al*. 2001) and Brazil (Resende *et al*. 2016) have made it possible to study the impact of the cultivation of transgenic corn on ***non-target insects*** (non targeted by the toxin). They also evaluated the environmental benefit provided by this *Bt* corn compared to the use of conventional chemical or biological insecticides. The results confirmed that the *Bt* varieties, due to their specificity to Lepidoptera (especially the diamondback moth, *Plutella xylostella*), respect non-target insects and in particular those that are beneficial for crops, unlike the use of a chemical active substance, which acts as systemic insecticide.

The drawback of this strategy, however, is that insects can quickly adapt to toxins. Tabashnik *et al*. (2013) defined ***field-evolved (or field-selected) resistance*** as a genetically based decrease in susceptibility of an insect population to a toxin that is caused by exposure of the population to the toxin in the field. Recognizing that field-evolved resistance is not 'all or none', Tabashnik *et al*. (2014) described at least three categories of field-evolved resistance, which were reported in field and entail, each, a genetically based decrease in susceptibility to a toxin in one or more field populations:

a. **Incipient resistance**, <1% resistant individuals.
b. **Early warning of resistance**, 1 to 6% resistant individuals;
c. **Practical resistance**, >50% resistant individuals and reduced efficacy reported

1.6. RNA-Based Antiviral Resistance

The role of RNA transcripts of viral transgenes made a major contribution to the discovery of an entirely new field in biology involving sequence-specific RNA breakdown. This process, occurring in plants but also in other eukaryotes, is known as ***RNA interference*** (RNAi) or ***RNA silencing*** (Ding and Voinnet 2007). The process is based on the idea that sequence-specific RNA degradation induced by transgenes will target all RNAs with sequence identity with the transgene RNA. The basis of sequence-specific recognition was found to be determined by the generation of small interfering RNA (siRNA) molecules derived from the transgene (Hamilton and Baulcombe 1999). As these siRNA molecules were also observed in wild-type plants infected with viruses and viroids, it was concluded that the transgenic approach put into action a pre-existing antiviral defense in plants (Baulcombe 1996). RNA silencing functions as a natural immunity mechanism in plant defense against viral invasion. Hence, many viruses have evolved to express VSR proteins[5] to counter host antiviral RNA silencing (Burgyán and Havelda 2011). VSR-deficient mutant viruses in plant host cells are able to suppress ***RNA-based viral immunity (RVI)*** in plants defective in RNA silencing (Li and Ding 2006).

In their study, Niu *et al.* (2006), expressed amiRNAs (based on an *A. thaliana* miR159 precursor) targeting the sequence of two VSRs, P69 of the turnip yellow mosaic virus (TYMV) and HC-Pro of the turnip mosaic virus (TuMV), in *Arabidopsis*. As expected, transgenic plants expressing these two amiRNAs displayed specific resistance to TYMV and TuMV, indicating that the strategy was applicable in engineering antiviral plants.

Evidence, also, exists that RNA silencing also contributes to plant immunity against non-viral pathogens, as *R* proteins of the same NB-LRR family are able to perceive both viral and non-viral avirulence (*Avr*) effectors (Zvereva and Pooggin 2012).

2. PLANT GENOME EDITING

A new technique for genetic improvement of plants is starting to be widely used in laboratories around the world. It is the editing of genes by the Clustered Regularly Interspaced Short Palindromic Repeats, CRISPR-Cas9 (Ali *et al.* 2015; Chandrasekaran *et al.* 2016). Compared to previous techniques, this system makes it possible to target the location of the genome where we wish to introduce a modification, by means of a transgene, which does not integrate into the genome and which disappears during the following generations. The ***edited plants*** thus produced transmit the genetic modification to their descendants and show no genetic trace of this intervention. They are identical to plants with a natural variation in one of their genes. Therefore, Plant genome editing by CRISPR-Cas9 has been inspired by the allelic diversity of wild ancestors of cultivated species.

The CRISPR-Cas9 approach also makes it possible to target several genes simultaneously in a single operation, which allows considering the modification of multigene characters. Such an edition of multiple loci offers the possibility of improving the resistance of plants to parasites or to extreme climatic conditions, by modifying certain essential traits. For example, Wang *et al.* (2014) introduced targeted mutations in the three homoeoalleles that encode mildew-resistance locus (*mlo*) proteins, in wheat. The authors demonstrated that the mutagenesis of all three homoeologs in the same plant confers heritable broad-spectrum resistance to powdery mildew.

2.1. Modifying Host Plant Susceptibility Genes By Site-Directed-Mutagenesis and CRISPR-Cas9

The modification or removal of host susceptibility genes may be an effective strategy to achieve resistance by preventing their manipulation by the pathogens. For exemple, Cavatorta *et al.* (2011) demonstrated that transgenic overexpression of a site-directed-mutagenized viral-resistant allele of the eIF4E conferred resistance to Potato virus Y in potato. In a more recent study, Chandrasekaran *et al.* (2016) knocked out the eIF4E gene in cucumber using CRISPR-Cas and obtained homozygous mutant plants that are resistant to a variety of viral pathogens under greenhouse conditions. Susceptibility genes are often conserved among plant species (Huibers *et al.* 2013). Therefore, knowledge of susceptibility genes in one plant species can facilitate the discovery of new susceptibility genes in another plant species. The *Arabidopsis* susceptibility gene DOWNY MILDEW RESISTANT6 (DMR6) encodes an oxygenase, and its expression is up-regulated during pathogen infection (Van Damme *et al.* 2005, Van Damme *et al.* 2008; Zeilmaker *et al.* 2015). Modifying susceptibility genes to attenuate the virulence of pathogens is a strategy that holds great potential in crop protection. Nevertheless, it is important to evaluate the potential side-effects of modifying susceptibility genes, because a given allele that confers susceptibility to one pathogen may confer resistance to another or have other essential biological functions (Lorang *et al.* 2012; McGrann *et al.* 2014). It is also important to determine whether the observed resistance in laboratory experiments is robust and/or durable under field conditions.

CONCLUSION

Challenges are high concerning varietal resistance in crops, which offers a major opportunity to reduce the use of phytosanitary products. Since the first report of a genetically modified crop conferring disease resistance in 1986, there have been huge breakthroughs both in the laboratory and in the field. All of the new plant improvement techniques (mutagenesis, genome editing and transgenesis) are revolutionizing the current vision of agriculture and seeds. Several of these techniques have already made it possible to obtain improved cultivars carrying advantageous and stable characters.

[1] Refer to Chapter 19.

[2] Paradigm shifts occur when the dominant paradigm under which normal science operates becomes incongruent with new phenomena, enabling the adoption of a new theory or paradigm (Kuhn 1962).

[3] The first plant defensins Rs-AFP1 (44 aminoacids) and Rs-AFP2 (36 aminoacids) with antifungal activity were isolated from radish seeds (Terras *et al.* 1992).

[4] Fire blight is the major bacterial disease of Maloideae (pear, apple, and other members of the *Rosaceae*) caused by the necrogenic bacterium *Erwinia amylovora*.

[5]*i.e.* Viral suppressors of RNA-based viral immunity; Reviewed in Wu *et al.* (2010).

CHAPTER 21

Durability of Plant Resistance to Pathogens and Pests

Abstract: The methods of management of pathogens and pests have been changing during these years, under the pressure of current societal and political demands. To overcome the drawbacks of chemical control, it is possible to mobilize genetic (*i.e.* varieties resistant to diseases) and agronomic controlling methods (cultural practices favoring these resistances or reducing the risks of pressure or development of pests). However, the low durability of genetic resistance, which is linked to the adaptation of pathogens, imposes the need to propose solutions in order to improve the durability of genetic resistance. Resistance is said to be durable when its effectiveness lasts for many years in a large spatial environment, at high pressure from the pathogen, favoring, *a priori*, the selection of virulent variants. The combination of quantitative and qualitative resistance is among the best solutions, but the strategy for deploying the *R* genes is an interesting track to follow.

Keywords: Durability of plant resistance to pathogens, Gene pyramiding, Gene rotation, Quantitative resistance, Qualitative resistance, *R*-gene deployment strategy.

INTRODUCTION

The different attack and defense strategies implemented by plants and their pathogens result from a long co-evolution between interacting organisms and can be described according to the concept of the ***arms race***. Indeed, ***in natural ecosystems***, a population of pathogens having acquired the capacity to bypass the basal defenses of a given plant will be able to colonize it.

In standardized agrosystems, the slow coevolution between plants and their pathogens is greatly accelerated through agricultural practices. In fact, breeders generally favor the introgression of major resistance genes into varieties. However, these are cultivated within large homogeneous plots, with a low crop rotation. The selection pressure exerted on pathogens by cultivated plants is therefore much stronger than in natural ecosystems. Under these conditions, microorganisms have the capacity to adapt quickly to the selection pressures to which they are subjected. Numerous cases of monogenic resistance breakdown have been reported, among which we can cite the cases of powdery mildew resistance in cereals (Hovmøller *et al.* 2000), grapevine downy mildew resistance in *Vitis vinifera* (Peressotti *et al.* 2010), and tomato mosaic virus (ToMV) resistance in tomato (Lanfermeijer *et al.* 2005).

This permanent coevolution as well as the rapid adaptive molecular changes that it implies in pathogens, imposes to the plant breeder (geneticist) a *race against time* to constantly identify new sources of resistance. The transfer of efficient genes, into commercial cultivars, has been accelerated, thanks to molecular and genetic engineering technologies.

The question of the durability of resistance can be considered as a problem of adaptive response in populations of pathogens to the selection exercised by resistant hosts (McDonald and Linde 2002). Strategies to maximize sustainability should therefore both limit the selection of virulent pathogen genotypes and reduce the size of the pathogen populations (Mundt *et al.* 2002).

1. MECHANISMS FOR OVERCOMING SPECIFIC RESISTANCE

Durability of resistance is defined as the time between the introduction of the cultivar into the landscape and the time when the frequency of virulent pathotypes reaches a predefined threshold (van den Bosch and Gilligan 2003). When a resistance has been overcome, it loses its effectiveness following pathogen adaptation. Such an adaptation of pathogenic populations to the resistance gene, found in host plants, occurs when individuals virulent for this resistance become progressively more frequent. In other words, there is evolution of the genetic structure of pathogen populations due to selection forces, resulting in the resistance gene then becoming ineffective.

The circumvention of a specific resistance is dependent on the size and the genetic structure of the pathogen population, as well as on the pathogen biological characteristics such as its regime of reproduction. The parasite's ability to evolve is governed by the five classic evolutionary forces: mutation, genetic drift, migration, recombination and selection (McDonald and Linde 2002). Mutation and sexual recombination *generate genetic diversity (new alleles)* in pathogenic populations. In particular, the mutation is directly responsible for overcoming specific resistances. Genetic drift, selection and migration influence the *distribution of genetic diversity (change in allele frequencies)* (Table 1).

Table 1. Evolutionary forces involved in the adaptation of pathogens (adapted from Hossard *et al.* 2010).

Evolutionary Force	Mechanism of Action	Consequence
Mutation	(Random) change in the nucleotide sequence of certain genes	Birth of new alleles in the population
Genetic drift	Random fluctuation in the frequency of alleles in a population of limited numbers	Allele extinction or fixation
Migration	Flow of genes or genotypes between populations of a pathogen	Movement of virulent mutant genotypes between populations
Sexual recombination	Reassortment of alleles (*i.e.* Modification of allele associations)	Modification of global (multilocus) genetic diversity in the pathogen population
Selection	Higher or lower reproduction of a given genotype	Selection of the most successful pathogens

Interactions among these five evolutionary forces ultimately determine the genetic structure, and hence the evolutionary potential, of pathogen populations. These interactions can be explained through the following examples:

a. ***Example 1, interaction between mutation and selection:*** Mutation may produce mutant alleles at the avirulence locus of the pathogen resulting in a few virulent individuals, but if the plant doesn't exert a selection pressure on wild-type susceptible pathogen individuals, by means of its plant receptor (*R*-gene), then the virulent pathogen genotypes may never increase to a detectable frequency.

b. ***Example 2, interaction between mutation/selection and migration:*** Virulent pathogen mutants that originate and increase in frequency in a field containing a resistant cultivar may never cause a widespread epidemic if gene flow among fields is very low as a result of an effective quarantine.

c. ***Example 3, interaction between mutation/selection/migration and genetic drift:*** Highly virulent genotypes that are distributed over long distances may never become established in some locations because they experience extinctions as a result of genetic drift.

Nevertheless, according to McDonald and Linde (2002), ***selection remains the main evolutionary force***, which leads to overcoming resistance, *via* an increase in the frequency of mutant alleles, once the mutation for virulence has appeared. This selection of virulent pathotypes results in cycles of ***boom and bust*** (explosion and extinction). The cycle begins with the large-scale deployment of a single specific resistance gene (the boom). The population of the pathogen will then adapt to the presence of this new gene by evolving towards a new population capable of bypassing this gene (the bust): resistance is overcome and the resistant variety will no longer be used (Fig. 1). In this case, the selection pressure exerted by the massive use of a specific resistance gene has conferred a selective advantage on virulent individuals.

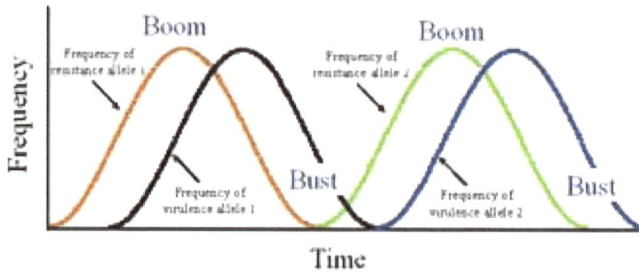

Fig. (1). Boom and bust phenomenon governing plant resistance gene overcoming by pathogens (credit to Aishakassimkassim1996, wikimedia commons).

This selection can occur more or less quickly after the introduction of the resistance gene depending on the pathogens, from one year for *Fusarium* wilt of tomato[1] to 8 years for apple scab[2] (McDonald and Linde 2002). This selection time can sometimes be very short: the example of rapeseed phoma illustrates the very rapid bypass of the specific resistance gene *Rlm1* (Rouxel *et al.* 2003)[3].

The durability of a resistance is conditioned by the biology of the pathogens, their dispersal capacities and the effective size of the populations. Generally, pathogens with a mixed mode of reproduction, with aerial dispersion and with large population sizes have the greatest (fastest) adaptive potential (McDonald and Linde 2002).

2. ASSOCIATION OF QUANTITATIVE AND QUALITATIVE RESISTANCE IN A SINGLE CULTIVAR

Monogenic resistance offers the advantage of being fully effective but has the disadvantage of being quickly overcome by pathogens and pests, which is less the case with partial resistances, called quantitative, for which the pathogen adaptation is more slow and difficult (Lindhout 2002; Stuthman *et al.* 2007). QTLs directly affect the evolution of the pathogen and slow the appearance of forms overcoming resistance (Quenouille *et al.* 2014). The mechanisms of action of QTLs are complex and understanding of the system requires sophisticated models[4]. The objective of these types of studies is to select plants with QTLs to protect a major resistance gene, which would then become more durable. Thus, the new varieties will be selected not only for their resistance, but also for their ability to control the evolution of pests over the long term.

The expected benefits of the combination of specific *R*-gene resistance and quantitative resistance in a single cultivar have been studied using mathematical models of the pathogen's evolution (Kiyosawa 1982; Pietravalle *et al.* 2006) but rarely confirmed experimentally. In 2009, Palloix *et al.*, using successive artificial reinoculations of viral isolates, showed the delayed emergence of virulent variants on hosts combining specific resistance (by *R* genes) and quantitative resistance compared to hosts with specific resistance by *R* genes alone (Palloix *et al.* 2009).

The combination of quantitative and qualitative resistance has proven to be an effective way to increase the durability of varietal resistance in the field, also in the *Leptosphaeria maculans*[5]-colza model (Delourme *et al.* 2006; Brun *et al.* 2010).

3. *R* GENE DEPLOYMENT STRATEGIES

Strategies for the resistance gene deployment (Fig. **2**) also influence the durability of varietal resistance. The massive and repeated deployment of a single source of resistance exerts a very strong selection pressure on pathogens and tends to favor the selection of virulent individuals. Conversely, the ***spatial and temporal alternation of resistance genes*** on a landscape scale can make it possible to increase the durability of resistance by maintaining diversity in populations.

Pyramiding resistance genes in varieties is also a possible approach to increase sustainability. Indeed, it can be intuitively postulated that the simultaneous overcoming of several major resistance genes requires the simultaneous establishment of mutations in separate genes and therefore results in too great fitness cost for pathogens (Mundt 2014).

3.1. One Gene at a Time

The durability of resistance necessarily depends on the genetic determinism of the interaction between the plant and the pathogen. Thus, a gene-for-gene monogenic interaction will lead more quickly to the circumvention of resistance by pathogens. There are some few counterexamples of major genes that have remained durable, such as the case of *mlo* resistance in barley. Long-term monogenic resistance is, however, quite rare in obligate parasitic fungi or in species with a narrow host range. There are a few cases of monogenic resistances that have never been overcome, such as those that control leaf rust (*Lr34* gene) and stem rust (*Sr31* gene) in wheat (Rahmatov *et al.* 2019). Also, the barley *mlo* gene confers a durable, broad-spectrum resistance to powdery mildew fungi in a multitude of plant species and its resistance has not been defeated (Kusch and Panstruga 2017)[6].

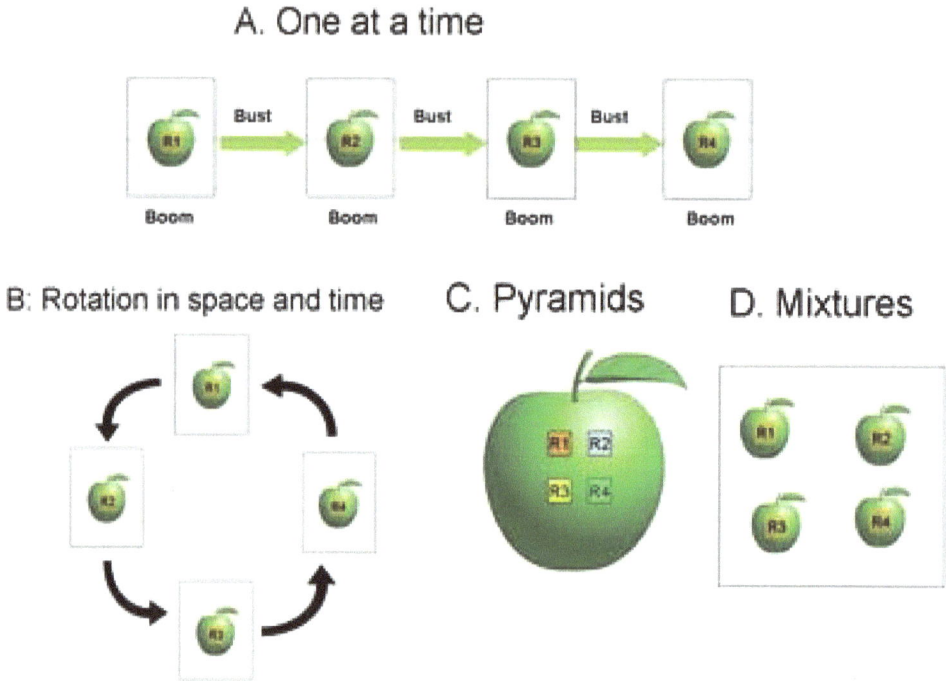

A. One at a time

B: Rotation in space and time   ## C. Pyramids   ## D. Mixtures

Fig. (2). Four strategies for deployment of major resistance genes (*R*-genes).
A. Deploying single *R*-genes one at a time in a sequence that matches the boom and bust cycle. **B. Rotations of *R*-genes in time or space.** Each *R*-gene is deployed over a limited number of years or area, and is withdrawn before the corresponding virulence allele achieves a high frequency in the pathogen population. **C. *R*-gene pyramid.** All *R*-genes are placed together in one plant genotype. **D. Cultivar mixtures.** Individual *R*-genes are grown as an intimate mixture in the same field.

3.2. Gene Rotation

Each *R*-gene is deployed over a limited number of years or area, and is withdrawn before the corresponding virulence allele achieves a high frequency in the pathogen population. This system of *R* gene management places longer selection pressure on the pathogen population (about twice as long a period as the traditional deployment of single *R* genes). The prolonged time, during which the selection pressure is exerted, along with the alternation of distinct selective pressures, together reduce the possibility of pathogenicity genes reaching high frequencies in the pathogen population.

3.3. Gene Pyramids

Root-knot nematodes (*Meloidogyne* spp.) are extremely polyphagous root pests with a very large impact at global scale. The strategy of resistance to the predominant species, *M. arenaria*, *M. incognita* and *M. javanica*, is a priority alternative in many cultivated plants, in particular in fruit trees of the genus *Prunus*. Three major dominant genes with different resistance spectra: *Ma* (plum), *RMia* (peach) and *RMja* (almond), have been identified and their pyramidization has been carried out in hybrids, paving the way for the creation of rootstocks protected by several genes simultaneously acting against each of the above-mentioned nematode species (Esmenjaud and Srinivasan 2012). This pyramiding guarantees the creation of plant material with unprecedented resistance durability (Duval *et al.* 2014).

We can also evoke, in this regard, the cases of pyramiding quantitative resistances, through the pyramidization of QRLs (Quantitative Resistance Loci). This has been revealed be more durable than monogenic resistances (French *et al.* 2016; Pilet-Nayel *et al.* 2017).

Various pyramidization strategies have been used in many species, and seem to be promising in terms of sustainability:

- Combination of QRLs with a major gene in a susceptible genetic background (Miklas *et al.* 2006; Palloix *et al.* 2009; Barbary *et al.* 2016),
- Accumulation of several major genes (Collard and Mackill 2008; Ellis *et al.* 2014),
- Association of several QRLs (Thabuis *et al.* 2004; Wilde *et al.* 2008).

Even though quantitative resistance is known to provide partial resistance, in some cases the accumulation of QRLs with strong effect may result in total resistance (Niks *et al.* 2015). An example is QRLs *rx1*, *rx2*, *rx3* for resistance of tomato to *Xanthomonas campestris* (Stall *et al.* 2009).

3.4. Cultivar Mixtures

Cultivar mixtures[7] refer to the simultaneous cultivation of several cultivars of the same species, leading to intraspecific diversity (Reiss and Drinkwater 2018). In some European countries, it has become common to cultivate mixtures of cultivars, that is, an equal mixture of seeds from 3 or 4 different cultivars, with different resistances (Jorgensen 1993). Mixtures of barley and oat cultivars are widely used in cereal-producing countries, such as Canada and Poland, mainly due to the great yield stability over the years and regions in addition to their simplicity and ease of implementation (Finckh *et al.* 2000). For example, in Poland, Tratwal and Bocianowski (2018) cited that about 17% of general growing area has been sown with mixtures (mainly of cereal and leguminous crops) in recent years. The same authors demonstrated that the use of a mixture made of five barley varieties ('Basza', 'Blask', 'Skarb', 'Rubinek', and 'Antek') significantly improved the control of powdery mildew and provided stable yield among different environments (Tratwal and Bocianowski 2018).

In the USA, blends composed of equal numbers of seeds of resistant and susceptible soybean lines were tested at three locations in Indiana and one location in Ohio, for three years, to assess yield losses attributable to *Phytophthora sojae*[8]. Results of this study revealed that plants of the resistant isoline were compensating for reduced productivity of the susceptible plants in the blend, showing the advantage of the cultivar mixture (Wilcox and St Martin 1998).

In China, nearly 90% of cultivar mixture cropping treatments experienced an increase in Winter Wheat yield (Wang *et al.* 2016b). In the same country, an advantage of cultivar mixtures has been also demonstrated in terms of containment of fungal diseases in rice (Zhu *et al.* 2000). Wheat cultivars mixture was also reported to restrict yellow and leaf rusts and foliar blight in a study conducted over a diverse range of environments in South Asia, including India, Pakistan and Nepal (Dubin and Wolfe 1994).

In Eastern Africa (Alupe, Western Kenya), Ngugi *et al.* (2001) used mixtures of sorghum to control anthracnose (caused by *Colletotrichum sublineolum*) and leaf blight (caused by *Exserohilum turcicum*), considered the most destructive diseases of high yielding cultivars in eastern Africa. A mixture made of one sorghum cultivar susceptible to both diseases and another with good resistance to both. Mixtures were found effective in controlling both diseases, delaying the time when the disease was first observed and lowering the rate of disease progress.

On a global scale, it is increasingly assumed that the effective lifespan of resistance is extended when cultivating mixtures of cultivars with different specific resistance genes. However, the main problem of mixtures perhaps lies in the search for the correct combination of varieties, in order to avoid the development of ***super-races of the pathogen***. Some negative effects of cultivar mixtures have rarely been reported. For instance, there was no advantage in barley cultivars mixtures in a study conducted by Baker and Briggs (1984).

CONCLUSION

Cultivar resistance has become an important component of the system of agricultural practices. However, the decline in the use of phytosanitary products can be accompanied by the increase in certain diseases, usually controlled by chemical treatments against various pathogens. In this case, *a smart genetic management* must be considered with the aim of ensuring the effectiveness of cultivar resistance over time (*i.e.* durability).

[1] Caused by *Fusarium oxysporum* f. sp. *Lycopersici*; a devastating disease in major tomato- growing regions worldwide.

[2] Caused by the ascomycete fungus *Venturia inaequalis*

[3] The example of the *Rlm1* gene is a typical illustration of a boom-and-bust cycle and the adaptive potential of aerial parasites in field crops. *Rlm1* had been introduced into rapeseed varieties in the early 1990s in France, in a context where more than 90% of phoma strains carried the avirulent allele of the corresponding gene, *AvrLm1* (Rouxel *et al.* 2003). This gene has been shown to be very effective. It was therefore widely used, covering up to 45% of French rapeseed areas in 1999 (Rouxel *et al.* 2003). This gene thus exerted a very strong pressure on the fungus, leading to a rapid selection of virulent strains in the populations of phoma. This has resulted in overcoming *Rlm1* resistance in a few years.

[4] For more information, see Chapter 9, Quantitative resistance, paragraph: **1. Molecular mechanisms associated with quantitative immunity**

[5] Fungal agent of oilseed rape (*Brassica napus*) phoma disease.

[6]*mlo* has been described by Kusch and Panstruga (2017) as a **reliable, universal and effective weapon, to protect plants from infection by powdery mildew fungi**.

[7] The terms Variety Mixtures and Multilines can also be used almost synonymously.

[8]*Phytophthora sojae* is an oomycete and a soil-borne plant pathogen that causes stem and root rot of soybean.

References

Abbas, MST (2018) Genetically engineered (modified) crops (*Bacillus thuringiensis* crops) and the world controversy on their safety. *Egypt J Biol Pest Control,* 28, 52.
[http://dx.doi.org/10.1186/s41938-018-0051-2]

Acevedo-Garcia, J, Kusch, S & Panstruga, R (2014) Magical mystery tour: MLO proteins in plant immunity and beyond. *New Phytol,* 204, 273-81.
[http://dx.doi.org/10.1111/nph.12889]

Agrawal, GK, Rakwal, R, Jwa, NS & Agrawal, VP (2001) Signalling molecules and blast pathogen attack activates rice *OsPR1a* and *OsPR1b* genes: A model illustrating components participating during defense/stress response. *Plant Physiol Biochem,* 39, 1095-103.
[http://dx.doi.org/10.1016/S0981-9428(01)01333-X]

Aharoni, A & Vorst, O (2002) DNA microarrays for functional plant genomics. *Plant Mol Biol,* 48, 99-118.
[http://dx.doi.org/10.1023/a:1013734019946]

Albrecht, V, Ritz, O, Linder, S, Harter, K & Kudla, J (2001) The NAF domain defines a novel protein-protein interaction module conserved in Ca^{2+}-regulated kinases. *EMBO J,* 20, 1051-63.
[http://dx.doi.org/10.1093/emboj/20.5.1051]

Ali, S, Ganai, BA, Kamili, AN, Bhat, AA, Mir, ZA, Bhat, JA, Tyagi, A, Islam, ST, Mushtaq, M, Yadav, P, Rawat, S & Grovera, A (2018) Pathogenesis-related proteins and peptides as promising tools for engineering plants with multiple stress tolerance. *Microbiol Res,* 212–213, 29-37.
[http://dx.doi.org/10.1016/j.micres.2018.04.008]

Ali, Z, Abul-faraj, A, Li, L, Ghosh, N, Piatek, M, Mahjoub, A, Aouida, M, Piatek, A, Baltes, NJ, Voytas, DF, Dinesh-Kumar, S & Mahfouz, MM (2015) Efficient Virus-Mediated Genome Editing in Plants Using the CRISPR/Cas9 System. *Mol Plant,* 8, 1288-91.
[http://dx.doi.org/10.1016/j.molp.2015.02.011]

Altpeter, F, Diaz, I, McAuslane, H, Gaddour, K, Carbonero, P & Vasil, IK (1999) Increased insect resistance in transgenic wheat stably expressing trypsin inhibitor CMe. *Mol Breed,* 5, 53-63.
[http://dx.doi.org/10.1023/A:1009659911798]

Alves, MS, Soares, ZG, Vidigal, PMP, Barros, EG, Poddanosqui, AMP, Aoyagi, LN, Abdelnoor, RV, Marcelino-Gumaraes, FC & Fietto, LG (2015) Differential expression of four soybean bZIP genes during *Phakopsora pachyrhizi* infection. *Funct Integr Genomics,* 15, 685-96.
[http://dx.doi.org/10.1007/s10142-015-0445-0]

Anderson, CM, Wagner, TA, Perret, M, He, ZH, He, D & Kohorn, BD (2001) WAKs: cell wall-associated kinases linking the cytoplasm to the extracellular matrix. *Plant Mol Biol,* 47, 197-206.

Andolfo, G, Donato, AD, Chiaiese, P, Natale, AD, Pollio, A, Jones, JDG, Frusciante, L & Ercolano, MR (2019) Alien domains shaped the modular structure of plant NLR proteins. *Genome Biol Evol,* 11, 3466-77.
[http://dx.doi.org/10.1093/gbe/evz248]

Andreu, AB, Guevara, MG, Wolski, EA, Daleo, GR & Caldiz, DO (2006) Enhancement of natural disease resistance in potatoes by chemicals. *Pest Manag Sci,* 62, 162-70.
[http://dx.doi.org/10.1002/ps.1142]

Anuradha, TS, Divya, EK, Jami, ESK & Kirti, EPB (2008) Transgenic tobacco and peanut plants expressing a mustard defensin show resistance to fungal pathogens. *Plant Cell Rep,* 27, 1777-86.
[http://dx.doi.org/10.1007/s00299-008-0596-8]

Arabidopsis Genome Initiative (2000) Analysis of the genome sequence of the flowering plant Arabidopsis thaliana. *Nature,* 408, 796-815.
[http://dx.doi.org/10.1038/35048692]

Araújo, ACde, Fonseca, FCDA, Cotta, MG, Alves, GSC & Miller, RNG (2019) Plant NLR receptor proteins and their potential in the development of durable genetic resistance to biotic stresses. *Biotechnology Research and Innovation,* 3, 80-94.
[http://dx.doi.org/10.1016/j.biori.2020.01.002]

Arruda, RL, Paz, ATS & Bara, MTF (2016) An approach on phytoalexins: function, characterization and biosynthesis in plants of the family Poaceae. *Ciência Rural, Santa Maria,* 46, 1206-16.
[http://dx.doi.org/10.1590/0103-8478cr20151164]

Arya, P, Kumar, G, Acharya, V & Singh, AK (2014) Genome-Wide Identification and Expression Analysis of NBS-Encoding Genes in *Malus x domestica* and expansion of NBS genes family in Rosaceae. *PLoS One,* 9, e107987.
[http://dx.doi.org/10.1371/journal.pone.0107987]

Asai, T, Tena, G, Plotnikova, J, Willmann, MR & Chiu, W-L (2002) MAP Kinase Signalling Cascade in Arabidopsis Innate Immunity. *Nature,* 415, 977-83.
[http://dx.doi.org/10.1038/415977a]

Ashfield, T, Egan, AN & Pfeil, BE (2012) Evolution of a Complex Disease Resistance Gene Cluster in Diploid Phaseolus and Tetraploid Glycine. *Plant Physiol,* 159, 336-54.
[http://dx.doi.org/10.1104/pp.112.195040]

Ashrafi, H, Hill, T, Stoffel, K, Kozik, A, Yao, J, Chin-Wo, SR & Van Deynze, A (2012) *De novo* assembly of the pepper transcriptome (*Capsicum annuum*): a benchmark for *in silico* discovery of SNPs, SSRs and candidate genes. *BMC Genomics,* 13, 571.
[http://dx.doi.org/10.1186/1471-2164-13-571]

Atkinson, NJ (2011) Plant molecular response to combined drought and nematode stress. PhD thesis, University of Leeds. http://etheses.whiterose.ac.uk/2131/

Atkinson, NJ, Dew, TP, Orfila, C & Urwin, PE (2011) Influence of Combined Biotic and Abiotic Stress on Nutritional Quality Parameters in Tomato (*Solanum Lycopersicum*). *J Agric Food Chem,* 59, 9673-82.
[http://dx.doi.org/10.1021/jf202081t]

Aylward, J, Steenkamp, ET, Dreyer, LL, Roets, F, Wingfield, BD & Wingfield, MJ (2017) A plant pathology perspective of fungal genome sequencing. *IMA Fungus,* 8, 1-15.
[http://dx.doi.org/10.5598/imafungus.2017.08.01.01]

Aziz, A, Poinssot, B, Daire, X, Adrian, M, Bézier, A, Lambert, B & Joubert, JM (2003) Laminarin elicits defense responses in grapevine and induces protection against *Botrytis cinerea* and *Plasmopara viticola. Mol Plant Microbe Interact,* 16, 1118-28.
[http://dx.doi.org/10.1094/MPMI.2003.16.12.1118]

Baggs, E, Dagdas, G & Krasileva, K (2017) NLR diversity, helpers and integrated domains: making sense of the NLR IDentity. *Curr Opin Plant Biol,* 38, 59-67.

[http://dx.doi.org/10.1016/j.pbi.2017.04.012]

Bai, Y, Kissoudis, C, Yan, Z, Visser, RGF & van der Linden, G (2018) Plant behaviour under combined stress: tomato responses to combined salinity and pathogen stress. *Plant J*, 93, 781-93.
[http://dx.doi.org/10.1111/tpj.13800]

Baillo, EH, Kimotho, RN, Zhang, Z & Xu, P (2019) Transcription Factors Associated with Abiotic and Biotic Stress Tolerance and Their Potential for Crops Improvement. *Genes (Basel)*, 10, 771.
[http://dx.doi.org/10.3390/genes10100771]

Baker, RJ & Briggs, KG (1984) Comparison of grain-yield of uniblends and biblends of 10 spring barley cultivars. *Crop Sci*, 24, 85-7.

Bakker, PAHM, Doornbos, RF, Zamioudis, C, Berendsen, RL & Pieterse, CMJ (2013) Induced Systemic Resistance and the Rhizosphere Microbiome. *Plant Pathol J*, 29, 136-43.
[http://dx.doi.org/10.5423/PPJ.SI.07.2012.0111]

Balagué, C, Lin, B, Alcon, C, Flottes, G, Malmström, S, Köhler, C, Neuhaus, G, Pelletier, G, Gaymard, F & Roby, D (2003) HLM1, an Essential Signaling Component in the Hypersensitive Response, Is a Member of the Cyclic Nucleotide-Gated Channel Ion Channel Family. *Plant Cell*, 15, 365-79.
[http://dx.doi.org/10.1105/tpc.006999]

Balandín, M, Royo, J, Gómez, E, Muniz, LM, Molina, A & Hueros, G (2005) A protective role for the embryo surrounding region of the maize endosperm, as evidenced by the characterisation of *ZmESR*-6, a defensin gene specifically expressed in this region. *Plant Mol Biol*, 58, 269-82.
[http://dx.doi.org/10.1007/s11103-005-3479-1]

Balbi, V & Devoto, A (2008) Jasmonate signalling network in *Arabidopsis thaliana*: crucial regulatory nodes and new physiological scenarios. *New Phytol*, 177, 301-18.
[http://dx.doi.org/10.1111/j.1469-8137.2007.02292.x]

Ballini, E, Morel, JB, Droc, G, Price, A, Courtois, B, Notteghem, JL & Tharreau, D (2008) A genome-wide meta-analysis of rice blast resistance genes and quantitative trait loci provides new insights into partial and complete resistance. *Mol Plant Microbe Interact*, 21, 859-68.
[http://dx.doi.org/10.1094/mpmi-21-7-0859]

Banerjee, A & Roychoudhury, A (2015) WRKY Proteins: Signaling and Regulation of Expression during Abiotic Stress Responses. *ScientificWorldJournal*, 807560
[http://dx.doi.org/10.1155/2015/807560]

Barbary, A, Djian-Caporalino, C, Marteu, N, Fazari, A & Caromel, B (2016) Plant genetic background increasing the efficiency and durability of major resistance genes to root-knot nematodes can be resolved into a few resistance QTLs. *Front Plant Sci*, 7, 56-64.
[http://dx.doi.org/10.3389/fpls.2016.00632]

Barka, EA, Eullaffroy, P, Clément, C & Vernet, G (2004) Chitosan Improves Development, and Protects *Vitis Vinifera* L. Against *Botrytis Cinerea*. *Plant Cell Rep*, 22, 608-14.
[http://dx.doi.org/10.1007/s00299-003-0733-3]

Barreto-Bergter, E, Sassaki, GL & de Souza, LM (2011) Structural analysis of fungal cerebrosides. *Front Microbiol*, 2, 239.
[http://dx.doi.org/10.3389/fmicb.2011.00239]

Barrett, LG, Kniskern, JM, Bodenhausen, N, Zhang, W & Bergelson, J (2009) Continua of specificity and virulence in plant host–pathogen interactions: Causes and consequences. *New Phytol*, 183, 513-29.

[http://dx.doi.org/10.1111/j.1469-8137.2009.02927.x]

Bartoli, C & Roux, F (2017) Genome-wide association studies in plant pathosystems: toward an ecological genomics approach. *Front Plant Sci,* 8, 763.
[http://dx.doi.org/10.3389/fpls.2017.00763]

Baulcombe, D (1996) Mechanisms of pathogen-derived resistance to viruses in transgenic plants. *Plant Cell,* 8, 1833-44.
[http://dx.doi.org/10.1105/tpc.8.10.1833]

Bendahmane, A, Querci, M, Kanyuka, K & Baulcombe, D (2000) Agrobacterium transient expression system as a tool for the isolation of disease resistance genes: application to the Rx2 locus in potato. *Plant J,* 21, 73-81.
[http://dx.doi.org/10.1046/j.1365-313x.2000.00654.x]

Benhamou, N (2009) *La résistance chez les plantes: Principes de la stratégie défensive et applications agronomiques,* (Éditions TEC & DOC,), Lavoisier, Paris.

Benhamou, N & Picard, K (1999) La résistance induite: une nouvelle stratégie de défense des plantes contre les agents pathogènes. *Phytoprotection,* 80, 137-99. [Induced resistance: a novel plant defense strategy against pathogens].
[http://dx.doi.org/10.7202/706189ar]

Benhamou, N & Rey, P (2012) Stimulateurs des défenses naturelles des plantes: une nouvelle stratégie phytosanitaire dans un contexte d'écoproduction durable. II. Intérêt des SDN en protection des cultures. *Phytoprotection,* 92, 1-48. [Elicitors of natural plant defense mechanisms: a new management strategy in the context of sustainable production. II. Interest for SDN in crop protection].
[http://dx.doi.org/10.7202/1013299ar]

Bent, AF (1996) Plant disease resistance genes: Function meets structure. *Plant Cell,* 8, 1757-71.
[http://dx.doi.org/10.1105/tpc.8.10.1757]

Bent, AF, Kunkel, BN, Dahlbeck, D, Brown, KL, Schmidt, R, Giraudat, J, Leung, J & Staskawicz, BJ (1994) RPS2 of *Arabidopsis thaliana*: a leucine-rich repeat class of plant disease resistance genes. *Science,* 265, 1856-60.
[http://dx.doi.org/10.1126/science.8091210]

Bergelson, J, Kreitman, M, Stahl, EA & Tian, D (2001) Evolutionary dynamics of plant R genes. *Science,* 292, 2281-5.
[http://dx.doi.org/10.1126/science.1061337]

Berrocal-Lobo, M & Molina, A (2004) Ethylene Response Factor 1 Mediates Arabidopsis Resistance to the Soilborne Fungus Fusarium Oxysporum. *Mol Plant Microbe Interact,* 17, 763-70.
[http://dx.doi.org/10.1094/MPMI.2004.17.7.763]

Bettaieb, I & Bouktila, D (2020) Genome-wide analysis of NBS-encoding resistance genes in the Mediterranean olive tree (*Olea europaea* subsp. *europaea* var. *europaea*): insights into their molecular diversity, evolution and function. *Tree Genet Genomes,* 16, 23.
[http://dx.doi.org/10.1007/s11295-020-1415-9]

Bilichak, A & Kovalchuk, I (2016) Transgenerational response to stress in plants and its application for breeding. *J Exp Bot,* 67, 2081-92.
[http://dx.doi.org/10.1093/jxb/erw066]

Birch, PR, Boevink, PC, Gilroy, EM, Hein, I, Pritchard, L & Whisson, SC (2008) Oomycete RXLR effectors:

delivery, functional redundancy and durable disease resistance. *Curr Opin Plant Biol,* 11, 373-9.
[http://dx.doi.org/10.1016/j.pbi.2008.04.005]

Bizuneh, GK (2020) The chemical diversity and biological activities of phytoalexins. *Advances in Traditional Medicine,* 21, 31-43.
[http://dx.doi.org/10.1007/s13596-020-00442-w]

Blanchard, A (2007) Identification, polymorphisme et évolution moléculaire de gènes de pouvoir pathogène chez le nématode à kyste de la pomme de terre Globodera pallida. *Thèse Université de Rennes 1,* 230.https://tel.archives-ouvertes.fr/tel-00132028

Bleecker, AB & Kende, H (2000) Ethylene: a gaseous signal molecule in plants. *Annu Rev Cell Dev Biol,* 16, 1-18.
[http://dx.doi.org/10.1146/annurev.cellbio.16.1.1]

Blumwald, E, Aharon, GS & Lam, BCH (1998) Early signal transduction pathways in plant-pathogen interactions. *Trends Plant Sci,* 9, 342-6.
[http://dx.doi.org/10.1016/S1360-1385(98)01289-8]

Boeckler, GA, Gershenzon, J & Unsicker, SB (2011) Phenolic glycosides of the Salicaceae and their role as anti-herbivore defenses. *Phytochemistry,* 72, 1497-509.
[http://dx.doi.org/10.1016/j.phytochem.2011.01.038]

Boiteau, G & Vernon, RS (2001) Physical barriers for the control of insect pests.*Physical control methods in plant protection,* Springer-Verlag/INRA, Heidelberg, Germany 224-47.

Boller, T & Felix, G (2009) A Renaissance of Elicitors: Perception of Microbe-Associated Molecular Patterns and Danger Signals by Pattern-Recognition Receptors. *Annu Rev Plant Biol,* 60, 379-406.
[http://dx.doi.org/10.1146/annurev.arplant.57.032905.105346]

Boller, T & He, SY (2009) Innate immunity in plants: an arms race between pattern recognition receptors in plants and effectors in microbial pathogens. *Science,* 324, 742-4.
[http://dx.doi.org/10.1126/science.1171647]

Bonardi, V, Tang, S, Stallmann, A, Roberts, M, Cherkis, K & Dangl, JL (2011) Expanded functions for a family of plant intracellular immune receptors beyond specific recognition of pathogen effectors. *Proc Natl Acad Sci USA,* 108, 16463-8.
[http://dx.doi.org/10.1073/pnas.1113726108]

Bond, DM & Baulcombe, DC (2015) Epigenetic transitions leading to heritable, RNA-mediated *de novo* silencing in *Arabidopsis thaliana. Proc Natl Acad Sci USA,* 112, 917.
[http://dx.doi.org/10.1073/pnas.1413053112]

Bonde, MR, Micales, JA & Peterson, GL (1993) The Use of Isozyme Analysis for Identification of Plant-Pathogenic Fungi. *Plant Dis,* 77, 961-8.
[http://dx.doi.org/10.1094/PD-77-0961]

Borges, F & Martienssen, RA (2015) The Expanding World of Small RNAs in Plants. *Nat Rev Mol Cell Biol,* 16, 727-41.
[http://dx.doi.org/10.1038/nrm4085]

Borras-Hidalgo, O, Caprari, C, Hernandez-Estevez, I, Lorenzo, GD & Cervone, F (2012) A gene for plant protection: expression of a bean polygalacturonase inhibitor in tobacco confers a strong resistance against *Rhizoctonia solani* and two oomycetes. *Front Plant Sci,* 3, 268.

[http://dx.doi.org/10.3389/fpls.2012.00268]

Boschi, F, Schvartzman, C & Murchio, S (2017) Enhanced Bacterial Wilt Resistance in Potato through expression of Arabidopsis EFR and introgression of quantitative resistance from Solanum commersonii. *Front Plant Sci,* 8, 1642.
[http://dx.doi.org/10.3389/fpls.2017.01642]

Bostock, R, Pye, M & Roubtsova, T (2014) Predisposition in plant disease: exploiting the nexus in abiotic and biotic stress perception and response. *Annu Rev Phytopathol,* 52, 517-49.
[http://dx.doi.org/10.1146/annurev-phyto-081211-172902]

Botella, MA, Parker, JE, Frost, LN, Bittner-Eddy, PD, Beynon, JL, Daniels, MJ, Holub, EB & Jones, JDG (1998) Three genes of the *Arabidopsis* RPP1 complex resistance locus recognize distinct *Peronospora parasitica* avirulence determinants. *Plant Cell,* 10, 1847-60.
[http://dx.doi.org/10.1105/tpc.10.11.1847]

Botstein, D, White, R, Skolnick, M & Davis, RW (1980) Construction of a genetic map in man using restriction fragment length polymorphisms. *Am J Hum Genet,* 32, 314-31.

Bourguet, D, Chaufaux, J, Micoud, A, Delos, M, Naibo, B, Bombarde, F, Marque, G, Eychenne, N & Pagliari, C (2002) *Ostrinia Nubilalis* parasitism and the field abundance of non-target insects in transgenic *Bacillus thuringiensis* corn (*Zea mays*). *Environ Biosafety Res,* 1, 49-60.
[http://dx.doi.org/10.1051/ebr:2002005]

Bourque, G, Leong, B, Vega, VB, Chen, X, Lee, YL, Srinivasan, KG, Chew, JL, Ruan, Y, Wei, C-L, Ng, HH & Liu, ET (2008) Evolution of the mammalian transcription factor binding repertoire *via* transposable elements. *Genome Res,* 18, 1752-62.
[http://dx.doi.org/10.1101/gr.080663.108]

Boyes, DC, Nam, J & Dangl, JF (1998) The *Arabidopsis thaliana RPM1* disease resistance gene product is a periferal plasma membrane protein that is degraded coincident with the hypersensitive response. *Proc Natl Acad Sci USA,* 95, 15849-54.
[http://dx.doi.org/10.1073/pnas.95.26.15849]

Brenchley, R, Spannagl, M, Pfeifer, M, Barker, GL, D'Amore, R, Allen, AM, McKenzie, N, Kramer, M, Kerhornou, A & Bolser, D (2012) Analysis of the bread wheat genome using wholegenome shotgun sequencing. *Nature,* 491, 705-10.
[http://dx.doi.org/10.1038/nature11650]

Britten, R (2006) Transposable elements have contributed to thousands of human proteins. *Proc Natl Acad Sci USA,* 103, 1798-803.
[http://dx.doi.org/10.1073/pnas.0510007103]

Broekaert, WF, Delaure, SL, De Bolle, MF & Cammue, BP (2006) The role of ethylene in host-pathogen interactions. *Annu Rev Phytopathol,* 44, 393-416.
[http://dx.doi.org/10.1146/annurev.phyto.44.070505.143440]

Broglie, KE, Butler, KH, De Silva, CA, Frey, TJ, Hawk, JA, Multani, DS, Wolters, C & Petra, JC (2006) *Polynucleotides and Methods for Making Plants Resistant to Fungal Pathogens,* E.I. du Pont de Nemours and Company, Pioneer Hi-Bred International, Inc., University of Delaware, United States. Patent 20060223102.
[https://patents.google.com/patent/US20060223102].

Brown, JKM (2015) Durable resistance of crops to disease: a darwinian perspective. *Annual Review of Phytopathology,* 53, 513-39.

[http://dx.doi.org/10.1146/annurev-phyto-102313-045914]

Brown, JKM & Rant, JC (2013) Fitness costs and trade-offs of disease resistance and their consequences for breeding arable crops. *Plant Pathol,* 62, 83-95.
[http://dx.doi.org/10.1111/ppa.12163]

Brun, H, Chèvre, A-M, Fitt, BD, Powers, S, Besnard, A-L, Ermel, M, Huteau, V, Marquer, B, Eber, F & Renard, M (2010) Quantitative resistance increases the durability of qualitative resistance to *Leptosphaeria maculans* in *Brassica napus. New Phytol,* 185, 285-99.
[http://dx.doi.org/10.1111/j.1469-8137.2009.03049.x]

Brutus, A, Sicilia, F, Macone, A, Cervone, F & De Lorenzo, G (2010) A domain swap approach reveals a role of the plant wall-associated kinase 1 (WAK1) as a receptor of oligogalacturonides. *Proc Natl Acad Sci USA,* 107, 9452-7.
[http://dx.doi.org/10.1073/pnas.1000675107]

Burgyán, J & Havelda, Z (2011) Viral suppressors of RNA silencing. *Trends Plant Sci,* 16, 265-72.
[http://dx.doi.org/10.1016/j.tplants.2011.02.010]

Bürstenbinder, K, Möller, B, Plötner, R, Stamm, G, Hause, G, Mitra, D & Abel, S (2017) The IQD Family of Calmodulin-Binding Proteins Links Calcium Signaling to Microtubules, Membrane Subdomains, and the Nucleus. *Plant Physiol,* 173, 1692-708.
[http://dx.doi.org/10.1104/pp.16.01743]

Butterbach, P, Verlaan, MG, Dullemans, A, Lohuis, D, Visser, RGF, Bai, Y & Kormelink, R (2014) Tomato yellow leaf curl virus resistance by Ty-1 involves increased cytosine methylation of viral genomes and is compromised by cucumber mosaic virus infection. *Proc Natl Acad Sci USA,* 111, 12942-7.
[http://dx.doi.org/10.1073/pnas.1400894111]

Cacciola, SO, Bertaccini, A, Pane, A & Furneri, PM (2017) Spiroplasma spp.: A Plant, Arthropod, Animal and Human Pathogen.*Citrus Pathology,* IntechOpen, Rijeka, Croatia 31-51.

Caffier, V, Le Cam, B, Al Rifai, M, Bellanger, MN, Comby, M & Denance, C (2016) Slow erosion of a quantitative apple resistance to *Venturia inaequalis* based on an isolate-specific quantitative trait Locus. *Infect Genet Evol,* 44, 541-8.
[http://dx.doi.org/10.1016/j.meegid]

Caldwell, DG, McCallum, N, Shaw, P, Muehlbauer, GJ, Marshall, DF & Waugh, R (2004) A structured mutant population for forward and reverse genetics in barley (*Hordeum vulgare* L.). *Plant J,* 40, 143-50.
[http://dx.doi.org/10.1111/j.1365-313X.2004.02190.x]

Calenge, F & Durel, CE (2006) Both stable and unstable QTLs for resistance to powdery mildew are detected in apple after four years of field assessments. *Mol Breed,* 17, 329-39.
[http://dx.doi.org/10.1007/s11032-006-9004-7]

Calenge, F, Drouet, D, Denance, C, Van de Weg, WE, Brisset, MN & Paulin, JP (2005) Identification of a major QTL together with several minor additive or epistatic QTLs for resistance to fire blight in apple in two related progenies. *Theor Appl Genet,* 111, 128-35.
[http://dx.doi.org/10.1007/s00122-005-2002-z]

Calenge, F, Faure, A, Goerre, M, Gebhardt, C, Van de Weg, WE & Parisi, L (2004) Quantitative trait loci (QTL) analysis reveals both broad-spectrum and isolate-specific QTL for scab resistance in an apple progeny challenged with eight isolates of Venturia inaequalis. *Phytopathology,* 94, 370-9.
[http://dx.doi.org/10.1094/phyto.2004.94.4.370]

Cannon, SB, Zhu, H, Baumgarten, AM, Spangler, R, May, G, Cook, DR & Young, ND (2002) Diversity, distribution and ancient taxonomic relationships within the TIR and NonTIR NBS-LRR resistance gene subfamilies. *J Mol Evol,* 54, 548-62.
[http://dx.doi.org/10.1007/s0023901-0057-2]

Caplan, JL, Mamillapalli, P, Burch-Smith, TM, Czymmek, K & Dinesh-Kumar, SP (2008) Chloroplastic Protein NRIP1 Mediates Innate Immune Receptor Recognition of a Viral Effector. *Cell,* 132, 449-62.
[http://dx.doi.org/10.1016/j.cell.2007.12.031]

Caranta, C, Lefebvre, V & Palloix, A (1997) Polygenic resistance of pepper to potyviruses consists of a combination of isolate-specific and broad-spectrum quantitative trait loci. *Mol Plant Microbe Interact,* 10, 872-8.
[http://dx.doi.org/10.1094/mpmi.1997.10.7.872]

Cardoso-Silva, CB, Costa, EA, Mancini, MC, Balsalobre, TWA, Canesin, LEC & Pinto, LR (2014) *De novo* assembly and transcriptome analysis of contrasting sugarcane varieties. *PLoS One,* 9, e88462.
[http://dx.doi.org/10.1371/journal.pone.0088462]

Cassel, SL, Joly, S & Sutterwala, FS (2009) The NLRP3 inflammasome: a sensor of immune danger signals. *Semin Immunol,* 21, 194.
[http://dx.doi.org/10.1016/j.smim.2009.05.002]

Cavatorta, J, Perez, KW, Gray, SM, VanEck, J, Yeam, I & Jahn, M (2011) Engineering virus resistance using a modified potato gene. *Plant Biotechnol J,* 9, 1014-21.
[http://dx.doi.org/10.1111/j.1467-7652.2011.00622.x]

Cesari, S, Bernoux, M, Moncuquet, P, Kroj, T & Dodds, PN (2014a) A novel conserved mechanism for plant NLR protein pairs: the integrated decoy hypothesis. *Frontiers in Plant Science,* 5, 606.
[http://dx.doi.org/10.3389/fpls.2014.00606]

Cesari, S, Kanzaki, H, Fujiwara, T, Bernoux, M, Chalvon, V, Kawano, Y, Shimamoto, K, Dodds, P, Terauchi, R & Kroj, T (2014) The NB-LRR proteins RGA4 and RGA5 interact functionally and physically to confer disease resistance. *EMBO J,* 33, 1941-59. b
[http://dx.doi.org/10.15252/embj.201487923]

Cesari, S, Thilliez, G, Ribot, C, Chalvon, V, Michel, C, Jauneau, A, Rivas, S, Alaux, L, Kanzaki, H & Okuyama, Y (2013) The rice resistance protein pair rga4/rga5 recognizes the magnaporthe oryzae effectors AVR-Pia and AVR1-CO39 by direct binding. *The Plant Cell,* 25, 1463-81.
[http://dx.doi.org/10.1105/tpc.112.107201]

Chalal, M, Klinguer, A, Echairi, A, Meunier, P, Vervandier-Fasseur, D & Adrian, M (2014) Antimicrobial activity of resveratrol analogues. *Molecules,* 19, 7679-88.
[http://dx.doi.org/10.3390/molecules19067679]

Chandrasekaran, J, Brumin, M, Wolf, D, Leibman, D, Klap, C, Pearlsman, M, Sherman, A, Arazi, T & Gal-On, A (2016) Development of broad virus resistance in non-transgenic cucumber using CRISPR/Cas9 technology. *Mol Plant Pathol,* 17, 1140-53.
[http://dx.doi.org/10.1111/mpp.12375]

Chandrashekar, A & Satyanarayana, KV (2006) Disease and pest resistance in grains of sorghum and millets. *J Cereal Sci,* 44, 287-304.
[http://dx.doi.org/10.1016/j.jcs.2006.08.010]

Chang, C & Bleecker, AB (2004) Ethylene biology. More than a gas. *Plant Physiol,* 136, 2895-9.

[http://dx.doi.org/10.1104/pp.104.900122]

Chang, X, Heene, E, Qiao, F & Nick, P (2011) The Phytoalexin Resveratrol Regulates the Initiation of Hypersensitive Cell Death in *Vitis* Cell. *PLoS One,* 6, e26405.
[http://dx.doi.org/10.1371/journal.pone.0026405]

Charbonnier, E, Ronceux, A, Carpentier, AS, Soubelet, H & Barriuso, E (2015) *Pesticides: des impacts aux changements de pratiques,* Editions Quae, France 400.

Charmillot, P-J, Pasquier, D & Scalco, A (1998) Le virus de la granulose du carpocapse *Cydia pomonella*: efficacité en microparcelles, rémanence et rôle des adjuvants. *Rev Suisse Vitic Arboric Hortic,* 30, 61-4.

Chen, F, Dong, W & Zhang, J (2018) The Sequenced Angiosperm Genomes and Genome Databases. *Front Plant Sci,* 9, 418.
[http://dx.doi.org/10.3389/fpls.2018.00418]

Chen, ZJ, Sreedasyam, A & Ando, A (2020) Genomic diversifications of five *Gossypium* allopolyploid species and their impact on cotton improvement. *Nat Genet*
[http://dx.doi.org/10.1038/s41588-020-0614-5]

Cheng, W, Chiang, M & Hwang, S (2009) Antagonism between abscisic acid and ethylene in *Arabidopsis* acts in parallel with the reciprocal regulation of their metabolism and signaling pathways. *Plant Mol Biol,* 71, 61-80.
[http://dx.doi.org/10.1007/s11103-009-9509-7]

Chester, KS (1933) The Problem of Acquired Physiological Immunity in Plants. *Q Rev Biol,* 8, 129-54.
[http://dx.doi.org/10.1086/394430]

Chinchilla, D, Bauer, Z, Regenass, M, Boller, T & Felix, G (2006) The Arabidopsis receptor kinase FLS2 binds flg22 and determines the specificity of flagellin perception. *Plant Cell,* 18, 465-76.
[http://dx.doi.org/10.1105/tpc.105.036574]

Chinchilla, D, Shan, L, He, P, Vries, SD & Kemmerling, B (2009) One for all: the receptor-associated kinase BAK1. *Trends Plant Sci,* 14, 535-41.
[http://dx.doi.org/10.1016/j.tplants.2009.08.002]

Chithrashree, , Udayashankar, AC, Nayaka, SC, Reddy, MS & Srinivas, C (2011) Plant growth-promoting rhizobacteria mediate induced systemic resistance in rice against bacterial leaf blight caused by Xanthomonas oryzae pv. Oryzae. *Biological Control,* 59, 114-22.

Choi, J, Tanaka, K, Cao, Y, Qi, Y, Qiu, J, Liang, Y, Lee, SY & Stacey, G (2014) Identification of a plant receptor for extracellular ATP. *Science,* 343, 290-4.
[http://dx.doi.org/10.1126/science.343.6168.290]

Choi, M-S, Kim, W, Lee, C & Oh, C-S (2013) Harpins, Multifunctional Proteins Secreted by Gram-Negative Plant-Pathogenic Bacteria. *Mol Plant Microbe Interact,* 26, 1115-22.
[http://dx.doi.org/10.1094/MPMI-02-13-0050-CR]

Christopoulou, M, McHale, LK, Kozik, A, Wo, SR-C, Wroblewski, T & Michelmore, RW (2015) Dissection of Two Complex Clusters of Resistance Genes in Lettuce (Lactuca Sativa). *Mol Plant Microbe Interact,* 28, 751-65.
[http://dx.doi.org/10.1094/MPMI-06-14-0175-R]

Chu, Z, Yuan, M, Yao, J, Ge, X, Yuan, B, Xu, C, Li, X, Fu, B, Li, Z, Bennetzen, JL, Zhang, Q & Wang, S (2006) Promoter mutations of an essential gene for pollen development result in disease resistance in rice.

Genes Dev, 20, 1250-5.
[http://dx.doi.org/10.1101/gad.1416306]

Clough, SJ, Fengler, KA, Yu, IC, Lippok, B, Smith, RK, Jr & Bent, AF (2000) The Arabidopsis dnd1 Defense, No Death Gene Encodes a Mutated Cyclic Nucleotide-Gated Ion Channel. *Proc Natl Acad Sci USA,* 97, 9323-8.
[http://dx.doi.org/10.1073/pnas.150005697]

Cohn, J, Sessa, G & Martin, GB (2001) Innate Immunity in Plants. *Curr Opin Immunol,* 13, 55-62.
[http://dx.doi.org/10.1016/s0952-7915(00)00182-5]

Colilla, FJ, Rocher, A & Mendez, E (1990) Gamma-Purothionins: amino acid sequence of two polypeptides of a new family of thionins from wheat endosperm. *FEBS Lett,* 270, 191-4.
[http://dx.doi.org/10.1016/0014-5793(90)81265-P]

Collard, BCY & Mackill, DJ (2008) Marker-assisted selection: an approach for precision plant breeding in the twenty-first century. *Philos Trans R Soc Lond B Biol Sci,* 363, 557-72.
[http://dx.doi.org/10.1098/rstb.2007.2170]

Collins, A, Milbourne, D, Ramsay, L, Meyer, R, Chatot-Balandras, C, Oberhagemann, P, De Jong, W, Gebhardt, C, Bonnel, E & Waugh, R (1999) QTL for field resistance to late blight in potato are strongly correlated with maturity and vigour. *Mol Breed,* 5, 387-98.
[http://dx.doi.org/10.1023/A:1009601427062]

Cordero, JC & Skinner, DZ (2002) Isolation from alfalfa of resistance gene analogues containing nucleotide binding sites. *Theor Appl Genet,* 104, 1283-9.
[http://dx.doi.org/10.1007/s00122-001-0821-0]

Corneliussen, B, Holm, M, Waltersson, Y, Onions, J, Hallberg, B, Thornell, A & Grundström, T (1994) Calcium/calmodulin inhibition of basic-helix-loop-helix transcription factor domains. *Nature,* 368, 760-4.
[http://dx.doi.org/10.1038/368760a0]

Cortijo, S, Wardenaar, R, Colome-Tatche, M, Gilly, A, Etcheverry, M, Labadie, K, Caillieux, E, Hospital, F, Aury, J-M & Wincker, P (2014) Mapping the Epigenetic Basis of Complex Traits. *Science,* 343, 1145-8.
[http://dx.doi.org/10.1126/science.1248127]

Corwin, JA & Kliebenstein, DJ (2017) Quantitative resistance: more than just perception of a pathogen. *Plant Cell,* 29, 655-65.
[http://dx.doi.org/10.1105/tpc.16.00915]

Cowger, C & Mundt, CC (2002) Aggressiveness of Mycosphaerella graminicola isolates from susceptible and partially resistant wheat cultivars. *Phytopathology,* 92, 624-30.
[http://dx.doi.org/10.1094/phyto.2002.92.6.624]

Cragg, GM & Newman, DJ (2013) Natural products: a continuing source of novel drug leads. *Biochim Biophys Acta,* 1830, 3670-95.
[http://dx.doi.org/10.1016/j.bbagen.2013.02.008]

D'Angelo, C, Weinl, S, Batistic, O, Pandey, GK, Cheong, YH, Schültke, S, Albrecht, V, Ehlert, B, Schulz, B & Harter, K (2006) Alternative complex formation of the Ca-regulated protein kinase CIPK1 controls abscisic acid-dependent and independent stress responses in Arabidopsis. *Plant J,* 48, 857-72.
[http://dx.doi.org/10.1111/j.1365-313X.2006.02921.x]

D'Ovidio, R, Mattei, B, Roberti, S & Bellincampi, D (2004) Polygalacturonases, polygalacturonase-inhibiting proteins and pectic oligomers in plant pathogen interactions. *Biochimica et Biophysica Acta (BBA)-*

Proteins and Proteomics, 1696, 237-44.
[http://dx.doi.org/10.1016/j.bbapap.2003.08.012]

Danan, S, Veyrieras, JB & Lefebvre, V (2011) Construction of a potato consensus map and QTL meta-analysis offer new insights into the genetic architecture of late blight resistance and plant maturity traits. *BMC Plant Biol,* 11, 16.
[http://dx.doi.org/10.1186/1471-2229-11-16]

Danchin, E, Charmantier, A, Champagne, FA, Mesoudi, A, Pujol, B & Blanchet, S (2011) Beyond DNA: integrating inclusive inheritance into an extended theory of evolution. *Nat Rev Genet,* 12, 475-86.
[http://dx.doi.org/10.1038/nrg3028]

Dangl, JL, Horvath, DM & Staskawicz, BJ (2013) Pivoting the plant Immune system from dissection to deployment. *Science,* 341, 746-51.
[http://dx.doi.org/10.1126/science.1236011]

Dangl, JL & Jones, JDG (2001) Plant pathogens and integrated defence responses to infection. *Nature,* 411, 826-33.
[http://dx.doi.org/10.1038/35081161]

Holoch, D & Moazed, D (2015) RNA-mediated Epigenetic Regulation of Gene Expression. *Nat Rev Genet,* 16, 71-84.
[http://dx.doi.org/10.1038/nrg3863]

Danquah, A, de Zélicourt, A, Boudsocq, M, Neubauer, J, Frei Dit Frey, N, Leonhardt, N, Pateyron, S, Gwinner, F, Tamby, JP, Ortiz-Masia, D, Marcote, MJ, Hirt, H & Colcombet, J (2015) Identification and characterization of an ABA-activated MAP kinase cascade in *Arabidopsis thaliana. Plant J,* 82, 232-44.
[http://dx.doi.org/10.1111/tpj.12808]

Danquah, A, Zélicourt, A, Colcombet, J & Hirt, H (2014) The role of ABA and MAPK signaling pathways in plant abiotic stress responses. *Biotechnol Adv,* 32, 40-52.
[http://dx.doi.org/10.1016/j.biotechadv.2013.09.006]

Day, EH, Hua, X & Bromham, L (2016) Is specialization an evolutionary dead end? Testing for differences in speciation, extinction and trait transition rates across diverse phylogenies of specialists and generalists. *J Evol Biol,* 29, 1257-67.
[http://dx.doi.org/10.1111/jeb.12867]

De Vos, M, Denekamp, M, Dicke, M, Vuylsteke, M, Van Loon, LC, Smeekens, SCM & Pieterse, C (2006) The *Arabidopsis thaliana* transcription factor AtMYB102 functions in defense against the insect herbivore *Pieris rapae. Plant Signal Behav,* 1, 305-11.
[http://dx.doi.org/10.4161/psb.1.6.3512]

De Wit, PJ, Buurlage, MB & Hammond, KE (1986) The occurrence of host-, pathogen- and interaction-specific proteins in the apoplast of *Cladosporium fulvum* (syn. *Fulvia fulva*) infected tomato leaves. *Physiol Mol Plant Pathol,* 29, 159-72.
[http://dx.doi.org/10.1016/S0048-4059(86)80018-2]

de Zelicourt, A, Colcombet, J & Hirt, H (2016) The role of MAPK modules and ABA during abiotic stress signaling. *Trends Plant Sci,* 21, 677-85.
[http://dx.doi.org/10.1016/j.tplants.2016.04.004]

Delaney, TP, Uknes, S, Vernooij, B, Friedrich, L, Weymann, K, Negrotto, D, Gaffney, T, Gut-Rella, M, Kessmann, H, Ward, E & Ryals, J (1994) A central role of salicylic acid in plant disease resistance. *Science,* 266, 1247-50.

[http://dx.doi.org/10.1126/science.266.5188.1247]

Delmas, CEL, Fabre, F, Jolivet, J, Mazet, ID, Cervera, SR & Deliere, L (2016) Adaptation of a plant pathogen to partial host resistance: selection for greater aggressiveness in grapevine downy mildew. *Evol Appl,* 9, 709-25.
[http://dx.doi.org/10.1111/eva.12368]

Delourme, R, Chèvre, A-M, Brun, H, Rouxel, T, Balesdent, M-H, Dias, J, Salisbury, P, Renard, M & Rimmer, SR (2006) Major gene and polygenic resistance to *Leptosphaeria maculans* in oilseed rape (*Brassica napus*). *Eur J Plant Pathol,* 114, 41-52.
[http://dx.doi.org/10.1007/s10658-005-2108-9]

Dempsey, DA & Klessig, DF (2012) SOS – too many signals for systemic acquired resistance? *Trends Plant Sci,* 17, 538-45.
[http://dx.doi.org/10.1016/j.tplants.2012.05.011]

Denby, KJ, Kumar, P & Kliebenstein, DJ (2004) Identification of Botrytis cinerea susceptibility loci in *Arabidopsis thaliana. Plant J,* 38, 473-86.
[http://dx.doi.org/10.1111/j.0960-7412.2004.02059.x]

Denoux, C, Galletti, R, Mammarella, N, Gopalan, S, Werck, D, De Lorenzo, G, Ferrari, S, Ausubel, FM & Dewdney, J (2008) Activation of defense response pathways by OGs and Flg22 elicitors in Arabidopsis seedlings. *Mol Plant,* 1, 423-45.
[http://dx.doi.org/10.1093/mp/ssn019]

Desgroux, A, L'Anthoene, V, Roux-Duparque, M, Riviere, JP, Aubert, G & Tayeh, N (2016) Genome-wide association mapping of partial resistance to *Aphanomyces euteiches* in pea. *BMC Genomics,* 17, 124.
[http://dx.doi.org/10.1186/s12864-016-2429-4]

Deslandes, L, Olivier, J, Peeters, N, Feng, DX, Khounlotham, M, Boucher, C, Somssich, I, Genin, S & Marco, Y (2003) Physical interaction between RRS1-R, a protein conferring resistance to bacterial wilt, and PopP2, a type III effector targeted to the plant nucleus. *Proc Natl Acad Sci USA,* 100, 8024-9.
[http://dx.doi.org/10.1073/pnas.1230660100]

Deslandes, L, Olivier, J, Theulieres, F, Hirsch, J, Feng, DX, Bittner-Eddy, P, Beynon, J & Marco, Y (2002) Resistance to *Ralstonia solanacearum* in *Arabidopsis thaliana* is conferred by the recessive RRS1-R gene, a member of a novel family of resistance genes. *Proc Natl Acad Sci USA,* 99, 2404-9.
[http://dx.doi.org/10.1073/pnas.032485099]

Deslandes, L & Rivas, S (2012) Catch me if you can: bacterial effectors and plant targets. *Trends Plant Sci,* 17, 644-55.
[http://dx.doi.org/10.1016/j.tplants.2012.06.011]

Devi, EL, Kumar, S, Singh, TB, Sharma, SK, Beemrote, A, Devi, CP, Chongtham, SK, Singh, CH, Yumlembam, RA, Haribhushan, A, Prakash, N & Wani, SH (2017) Adaptation Strategies and Defence Mechanisms of Plants During Environmental Stress. In: Ghorbanpour, M., Varma, A., (Eds.), *Medicinal Plants and Environmental Challenges*Springer, Cham.
[http://dx.doi.org/110.1007/978-3-319-68717-9_20]

DeYoung, BJ & Innes, RW (2006) Plant NBS-LRR proteins in pathogen sensing and host defense. *Nat Immunol,* 7, 1243-9.
[http://dx.doi.org/10.1038/ni1410]

Dghim, F, Bouaziz, M, Mezghani, I, Boukhris, M & Neffati, M (2015) Laticifers identification and natural rubber characterization from the latex of *Periploca angustifolia* Labill. (Apocynaceae). Flora - Morphology,

Distribution. *Functional Ecology of Plants,* 217, 90-8.
[http://dx.doi.org/10.1016/j.flora.2015.09.006]

Di Matteo, A, Federici, L, Mattei, B, Salvi, G, Johnson, KA, Savino, C, De Lorenzo, G, Tsernoglou, D & Cervone, F (2003) The crystal structure of polygalacturonase-inhibiting protein (PGIP), a leucine-rich repeat protein involved in plant defense. *Proc Natl Acad Sci USA,* 100, 10124-8.
[http://dx.doi.org/10.1073/pnas.1733690100]

Dietz, KJ, Vogel, MO & Viehhauser, A (2010) AP2/EREBP transcription factors are part of gene regulatory networks and integrate metabolic, hormonal and environmental signals in stress acclimation and retrograde signalling. *Protoplasma,* 245, 3-14.
[http://dx.doi.org/10.1007/s00709-010-0142-8]

Dima, JB, Sequeiros, C & Zaritzky, N (2017) Chitosan from Marine Crustaceans: Production, Characterization and Applications.*Biological Activities and Application of Marine Polysaccharides,* IntechOpen, Rijeka, Croatia 39-56.

Ding, SW & Voinnet, O (2007) Antiviral immunity directed by small RNAs. *Cell,* 130, 413-26.
[http://dx.doi.org/10.1016/j.cell.2007.07.039]

Dixon, MS, Jones, DA, Keddie, JS, Thomas, CM, Harrison, K & Jones, JD (1996) The tomato *Cf-2* disease resistance locus comprises two functional genes encoding leucine-rich repeat proteins. *Cell,* 84, 451-9.
[http://dx.doi.org/10.1016/s0092-8674(00)81290-8]

Djamei, A, Schipper, K & Rabe, F (2011) Metabolic priming by a secreted fungal effector. *Nature,* 478, 395-8.
[http://dx.doi.org/10.1038/nature10454]

Dodds, PN, Lawrence, GJ, Catanzariti, AM, Teh, T, Wang, CIA, Ayliffe, MA, Kobe, B & Ellis, JG (2006) Direct protein interaction underlies gene-for-gene specificity and coevolution of the flax resistance genes and flax rust avirulence genes. *Proc Natl Acad Sci USA,* 103, 8888-93.
[http://dx.doi.org/10.1073/pnas.0602577103]

Dodds, PN, Lawrence, GJ & Ellis, JG (2000) Six amino acid changes confined to the Leucine-Rich Repeat B-strand/B-turn motif determine the difference between the *P* and *P2* rust resistance specificities in flax. *Plant Cell,* 13, 163-78.
[http://dx.doi.org/10.1105/tpc.13.1.163]

Dodds, PN & Rathjen, JP (2010) Plant Immunity: Towards an Integrated View of Plant-Pathogen Interactions. *Nat Rev Genet,* 11, 539-48.
[http://dx.doi.org/10.1038/nrg2812]

Doi, Y, Teranaka, M, Yora, K & Asuyama, H (1967) Mycoplasma or PLT, group-like micro-organisms found in the phloem elements of plants infected with mulberry dwarf, potato witches' broom, aster yellows, or Paulownia witches' broom. *Ann Phytopathol Soc Jpn,* 33, 259-66.
[http://dx.doi.org/10.3186/jjphytopath.33.259]

Dong, X (1998) SA, JA, ethylene, and disease resistance in plants. *Curr Opin Plant Biol,* 1, 316-23.
[http://dx.doi.org/10.1016/1369-5266(88)80053-0]

Dong, XN (2004) NPR1, all things considered. *Curr Opin Plant Biol,* 7, 547-52.
[http://dx.doi.org/10.1016/j.pbi.2004.07.005]

Dou, J, Kim, H, Li, Y, Padmakshan, D, Yue, F, Ralph, J & Vuorinen, T (2018) Structural Characterization of Lignins from Willow Bark and Wood. *J Agric Food Chem,* 66, 7294-300.

[http://dx.doi.org/10.1021/acs.jafc.8b02014]

Double, ML, Jarosz, AM, Fulbright, DW, Davelos Baines, A & MacDonald, WL (2018) Evaluation of Two Decades of *Cryphonectria parasitica* Hypovirus Introduction in an American Chestnut Stand in Wisconsin. *Phytopathology,* 108, 702-10.
[http://dx.doi.org/10.1094/PHYTO-10-17-0354-R]

Doughari, JH (2015) An Overview of Plant Immunity. *J Plant Pathol Microbiol,* 6, 1000322.
[http://dx.doi.org/10.4172/2157-7471.1000322]

Dowen, RH, Pelizzola, M, Schmitz, RJ, Lister, R, Dowen, JM, Nery, JR, Dixon, JE & Ecker, JR (2012) Widespread Dynamic DNA Methylation in Response to Biotic Stress. *Proc Natl Acad Sci USA,* 109, E2183-91.
[http://dx.doi.org/10.1073/pnas.1209329109]

Du, J, Verzaux, E, Chaparro-Garcia, A, Bijsterbosch, G, Keizer, LP, Zhou, J, Liebrand, TW, Xie, C, Govers, F & Robatzek, S (2015) Elicitin recognition confers enhanced resistance to *Phytophthora infestans* in potato. *Nat Plants,* 1, 15034.
[http://dx.doi.org/10.1038/nplants.2015.34]

Du, L & Chen, Z (2000) Identification of genes encoding receptor-like protein kinases as possible targets of pathogen- and salicylic acidinduced *WRKY* DNA-binding proteins in *Arabidopsis. Plant J,* 24, 837-47.
[http://dx.doi.org/10.1111/j.1365-313X.2000.00923.x]

Du, L & Poovaiah, BW (2004) A novel family of Ca^{2+}/calmodulin-binding proteins involved in transcriptional regulation: interaction with fsh/Ring3 class transcription activators. *Plant Mol Biol,* 54, 549-69.
[http://dx.doi.org/10.1023/B:PLAN.0000038269.98972.bb]

Duan, X, Li, X, Xue, Q, Abo-EI-Saad, M, Xu, D & Wu, R (1996) Transgenic rice plants harboring an introduced potato proteinase inhibitor II gene are insect resistant. *Nat Biotechnol,* 14, 494-8.
[http://dx.doi.org/10.1038/nbt0496-494]

Dubin, HJ & Wolfe, MS (1994) Comparative behavior of three wheat cultivars and their mixture in India, Nepal and Pakistan. *Field Crops Res,* 39, 71-83.
[http://dx.doi.org/10.1016/0378-4290(94)90010-8]

Duewell, P, Kono, H & Rayner, KJ (2010) NLRP3 inflammasomes are required for atherogenesis and activated by cholesterol crystals. *Nature,* 464, 1357.
[http://dx.doi.org/10.1038/nature08938]

Dunning, FM, Sun, W, Jansen, KL, Helft, L & Bent, AF (2007) Identification and mutational analysis of Arabidopsis FLS2 leucine-rich repeat domain residues that contribute to flagellin perception. *Plant Cell,* 19, 3297-313.
[http://dx.doi.org/10.1105/tpc.106.048801]

Durrant, WE & Dong, X (2004) Systemic acquired resistance. *Annu Rev Phytopathol,* 42, 185-209.
[http://dx.doi.org/10.1146/annurev.phyto.42.040803.140421]

Duval, H, Hoerter, M, Polidori, J, Confolent, C, Masse, M, Moretti, A, Van Ghelder, C & Esmenjaud, D (2014) High resolution mapping of the RMia gene for resistance to root-knot nematodes in peach. *Tree Genet Genomes,* 10, 297-306.
[http://dx.doi.org/10.1007/s11295-013-0683-z]

Dyrka, W, Lamacchia, M & Durrens, P (2014) Diversity and variability of NOD-like receptors in fungi.

Genome Biol Evol, 6, 3137-58.
[http://dx.doi.org/10.1093/gbe/evu251]

Eberhardt, M, Lee, C & Liu, R (2000) Antioxidant activity of fresh apples. *Nature,* 405, 903-4.
[http://dx.doi.org/10.1038/35016151]

Ebner, C, Hoffmann-Sommergruber, K & Breiteneder, H (2001) Plant food allergens homologous to pathogenesis-related proteins. *Allergy,* 56, 43-4.
[http://dx.doi.org/10.1034/j.1398-9995.2001.00913.x]

Eitas, TK & Dangl, JL (2010) NB-LRR proteins: pairs, pieces, perception, partners, and pathways. *Curr Opin Plant Biol,* 13, 472-7.
[http://dx.doi.org/10.1016/j.pbi.2010.04.007]

Elliger, CA, Wong, Y, Chan, BG & Waiss, AC (1981) Growth inhibitors in tomato (*Lycopersicon*) to tomato fruitworm (*Heliothis zea*). *J Chem Ecol,* 7, 753-8.

Ellis, J, Dodds, P & Pryor, T (2000) Structure, function and evolution of plant disease resistance genes. *Curr Opin Plant Biol,* 3, 278-84.
[http://dx.doi.org/10.1016/S1369-5266(00)00080-7]

Ellis, JG (2016) Integrated decoys and effector traps: how to catch a plant pathogen. *BMC Biol,* 14, 13.
[http://dx.doi.org/10.1186/s12915-016-0235-8]

Ellis, JG, Lagudah, ES, Spielmeyer, W & Dodds, PN (2014) The past, present and future of breeding rust resistant wheat. *Front Plant Sci,* 5, 641.
[http://dx.doi.org/10.3389/fpls.2014.00641]

Ellis, JG, Lawrence, GJ, Luck, JE & Dodds, PN (1999) Identification of regions in alleles of the flax rust resistance gene L that determine differences in gene-for-gene specificity. *Plant Cell,* 11, 495-506.
[http://dx.doi.org/10.1105/tpc.11.3.495]

Elmore, JM, Lin, ZJD & Coaker, G (2011) Plant NB-LRR signaling: upstreams and downstreams. *Curr Opin Plant Biol,* 14, 365-71.
[http://dx.doi.org/10.1016/j.pbi.2011.03.011]

Esmenjaud, D & Srinivasan, C (2012) Molecular breeding.*Genetics, genomics and breeding of stone fruits,* CRC press, New York 150-210.

Espinas, NA, Saze, H & Saijo, Y (2016) Epigenetic control of defense signaling and priming in plants. *Front Plant Sci,* 7, 1201.
[http://dx.doi.org/10.3389/fpls.2016.01201]

Esteller, M (2011) Non-coding RNAs in human disease. *Nat Rev Genet,* 12, 861-74.
[http://dx.doi.org/10.1038/nrg3074]

Eulgem, T (2005) Regulation of the *Arabidopsis* defense transcriptome. *Trends Plant Sci,* 10, 71-8.
[http://dx.doi.org/10.1016/j.tplants.2004.12.006]

Eulgem, T, Rushton, PJ, Robatzek, S & Somssich, IE (2000) The *WRKY* superfamily of plant transcription factors. *Trends Plant Sci,* 5, 199-206.
[http://dx.doi.org/10.1016/s1360-1385(00)01600-9]

Fatima, U & Senthil-Kumar, M (2015) Plant and pathogen nutrient acquisition strategies. *Frontiers in Plant Science,* 6, 750.

[http://dx.doi.org/10.3389/fpls.2015.00750]

Fedoroff, NV (2012) Transposable elements, epigenetics, and genome evolution. *Science,* 338, 758-67. [http://dx.doi.org/10.1126/science.338.6108.758]

Felix, G, Duran, JD, Volko, S & Boller, T (1999) Plants have a sensitive perception system for the most conserved domain of bacterial flagellin. *Plant J,* 18, 265-76. [http://dx.doi.org/10.1046/j.1365-313x.1999.00265.x]

Chen, F, Ludwiczuk, A, Wei, G, Chen, X, Crandall-Stotler, B & Bowman, JL (2018) Terpenoid secondary metabolites in bryophytes: chemical diversity, biosynthesis and biological functions. *Crit Rev Plant Sci,* 37, 210-31. [http://dx.doi.org/10.1080/07352689.2018.1482397]

Feng, S, Jacobsen, SE & Reik, W (2010) Epigenetic reprogramming in plant and animal development. *Science,* 330, 622-7. [http://dx.doi.org/10.1126/science.1190614]

Finckh, MR, Gacek, ES & Goyeau, H (2000) Cereal variety and species mixtures in practice, with emphasis on disease resistance. *Agronomie,* 20, 813-37.

Fletcher, J, Bender, C, Budowle, B, Cobb, WT, Gold, SE, Ishimaru, CA, Luster, D, Melcher, U, Murch, R, Scherm, H, Seem, RC, Sherwood, JL, Sobral, BW & Tolin, SA (2006) Plant Pathogen Forensics: Capabilities, Needs, and Recommendations. *Microbiol Mol Biol Rev,* 70, 450-71. [http://dx.doi.org/10.1128/MMBR.00022-05]

Flor, HH (1971) Current status of the gene-for-gene concept. *Annu Rev Phytopathol,* 9, 275-96.

Fluhr, R (2001) Sentinels of disease. Plant resistance genes. *Plant Physiol,* 127, 1367-74. [http://dx.doi.org/10.1104/pp.010763]

Fonseca, S, Chico, JM & Solano, R (2009) The jasmonate pathway: the ligand, the receptor and the core signalling module. *Curr Opin Plant Biol,* 12, 539-47. [http://dx.doi.org/10.1016/j.pbi.2009.07.013]

Foster, J & Fermin, G (2018) Origins and Evolution of Viruses. In: Tennant, P., Fermin, G., Foster, J., (Eds.), *Viruses: Molecular Biology, Host Interactions, and Applications to Biotechnology* Academic Press.

Foster, JE & Fermin, G (2018) *Origins and Evolution of Viruses,* Viruses 83-100.

Fox, SE, Preece, J, Kimbrel, JA, Marchini, GL, Sage, A & Youens-Clark, K (2013) Sequencing and *de novo* transcriptome assembly of *Brachypodium sylvaticum* (Poaceae). *Appl Plant Sci,* 1, 1200011. [http://dx.doi.org/10.3732/apps.1200011]

Fradin, E, Adb-El-Haliem, A, Masini, L, van den Berg, G, Joosten, M & Thomma, B (2011) Inter-family transfer of tomato Ve1 mediates Verticillium resistance in Arabidopsis. *Plant Physiol,* 156, 2255-65. [http://dx.doi.org/10.1104/pp.111.180067]

Franco, OL, Murad, AM, Leite, JR, Mendes, PAM, Prates, MV & Bloch, C (2006) Identification of a cowpea γ-thionin with bactericidal activity. *FEBS J,* 273, 3489-97. [http://dx.doi.org/10.1111/j.1742-4658.2006.05349.x]

Freeman, BC & Beattie, GA (2008) An overview of plant defenses against pathogens and herbivores. *The Plant Health Instructor.*

[http://dx.doi.org/10.1094/PHI-I-2008-0226-01]

French, E, Kim, BS & Iyer-Pascuzzi, AS (2016) Mechanisms of quantitative disease resistance in plants. *Semin Cell Dev Biol,* 56, 201-8.
[http://dx.doi.org/10.1016/j.semcdb.2016.05.015]

Fritig, B, Heitz, T & Legrand, M (1998) Antimicrobial proteins in induced plant defense. *Curr Opin Immunol,* 10, 16-22.
[http://dx.doi.org/10.1016/S0952-7915(98)80025-3]

Fu, D, Uauy, C, Distelfeld, A, Blechl, A, Epstein, L, Chen, X, Sela, H, Fahima, T & Dubcovsky, J (2009) A kinase-START gene confers temperature-dependent resistance to wheat stripe rust. *Science,* 323, 1357-60.
[http://dx.doi.org/10.1126/science.1166289]

Fu, J & Wang, S (2011) Insights into auxin signaling in plant-pathogen interactions. *Front Plant Sci,* 2, 74.
[http://dx.doi.org/10.3389/fpls.2011.00074]

Furci, L, Jain, R, Stassen, J, Berkowitz, O, Whelan, J, Roquis, D, Baillet, V, Colot, V, Johannes, F & Ton, J (2019) Identification and characterisation of hypomethylated DNA loci controlling quantitative resistance in *Arabidopsis. eLife,* 8, e40655.
[http://dx.doi.org/10.7554/eLife.40655]

Gabriel, DW & Rolfe, BG (1990) Working models of specific recognition in plant-microbe interaction. *Annu Rev Phytopathol,* 28, 365-91.
[http://dx.doi.org/10.1146/annurev.py.28.090190.002053]

Gabriëls, SH, Vossen, JH, Ekengren, SK, van Ooijen, G, Abd-El-Haliem, AM, van den Berg, GC, Rainey, DY, Martin, GB, Takken, FL, de Wit, PJ & Joosten, MH (2007) An NB-LRR protein required for HR signalling mediated by both extra- and intracellular resistance proteins. *Plant J,* 50, 14-28.
[http://dx.doi.org/10.1111/j.1365-313X.2007.03027.x]

Galindo-Trigo, S, Gray, JE & Smith, LM (2016) Conserved Roles of CrRLK1L Receptor-Like Kinases in Cell Expansion and Reproduction from Algae to Angiosperms. *Front Plant Sci,* 7, 1269.
[http://dx.doi.org/10.3389/fpls.2016.01269]

Gao, L (2019) Structure Analysis of a Pathogenesis-Related 10 Protein from Gardenia jasminoides. *IOP Conference Series: Earth and Environmental Science,* 242, 4.
[http://dx.doi.org/10.1088/1755-1315/242/4/042005]

Gao, Y, Wang, W, Zhang, T, Gong, Z, Zhao, H & Han, G-Z (2018) Out of Water: The Origin and Early Diversification of Plant R-Genes. *Plant Physiol,* 177, 82-9.
[http://dx.doi.org/10.1104/pp.18.00185]

García-Arenal, F & McDonald, BA (2003) An analysis of the durability of resistance to plant viruses. *Phytopathology,* 93, 941-52.
[http://dx.doi.org/10.1094/phyto.2003.93.8.941]

Garcia-Brugger, A, Lamotte, O, Vandelle, E, Bourque, S, Lecourieux, D, Poinssot, B, Wendehenne, D & Pugin, A (2006) Early signaling events induced by elicitors of plant defenses. *Mol Plant Microbe Interact,* 19, 711-24.
[http://dx.doi.org/10.1094/MPMI-19-0711]

Garnier, M, Foissac, X, Gaurivaud, P, Laigret, F, Renaudin, J, Saillard, C & Bové, J (2001) Mycoplasmas, plants, insect vectors: A matrimonial triangle. *C R Acad Sci III,* 324, 923-8.

[http://dx.doi.org/10.1016/S0764-4469(01)01372-5]

Gebhardt, C, Ballvora, A, Walkemeier, B, Oberhagemann, P & Schuler, K (2004) Assessing genetic potential in germplasm collections of crop plants by marker-trait association: a case study for potatoes with quantitative variation of resistance to late blight and maturity type. *Mol Breed,* 13, 93-102.
[http://dx.doi.org/10.1023/B:MOLB.0000012878.89855.df]

Gechev, TS & Hille, J (2005) Hydrogen peroxide as a signal controlling plant programmed cell death. *J Cell Biol,* 168, 17-20.
[http://dx.doi.org/10.1083/jcb.200409170]

Gianinazzi, S, Martin, JC & Vallée, JC (1970) Hypersensibilité aux virus, température et protéines solubles chez *Nicotiana xanthi* n.c. Apparition de nouvelles macromolécules lors de la répression de la synthèse virale. *Comptes Rendus de l'Académie des Sciences Paris,* 270, 2383-6.

Giannakopoulou, A, Steele, JFC, Segretin, ME, Bozkurt, TO & Zhou, J (2015) Tomato I2 immune receptor can be engineered to confer partial resistance to the oomycete *Phytophthora infestans* in addition to the fungus *Fusarium oxysporum. Mol Plant Microbe Interact,* 28, 1316-29.
[http://dx.doi.org/10.1094/MPMI-07-15-0147-R]

Gilbert, GS & Webb, CO (2006) Phylogenetic signal in plant pathogen-host range. *Proc Natl Acad Sci USA,* 104, 4979-83.
[http://dx.doi.org/10.1073pnas.0607968104]

Gill, US, Lee, S & Mysore, KS (2015) Host Versus Nonhost Resistance: Distinct wars with similar arsenals. *Phytopathology,* 105, 580-7.
[http://dx.doi.org/10.1094/PHYTO-11-14-0298-RVW]

Giorgetti, C (2013) Part relative de l'architecture et de la résistance partielle dans le contrôle génétique du ralentissement des épidémies d'ascochytose à Didymella pinodes chez le pois. PhD Thesis, Agrocampus Ouest, 183p. [in French].

Glazebrook, J (2005) Contrasting mechanisms of defense against biotrophic and necrotrophic pathogens. *Annu Rev Phytopathol,* 43, 205-27.
[http://dx.doi.org/10.1146/annurev.phyto.43.040204.135923]

Glowacki, S, Macioszek, VK & Kononowicz, AK (2011) *R* proteins as fundamentals of plant innate immunity. *Cell Mol Biol Lett,* 16, 1-24.
[http://dx.doi.org/10.2478/s11658-010-0024-2]

Goff, SA, Ricke, D, Lan, TH, Presting, G, Wang, R, Dunn, M, Glazebrook, J, Sessions, A, Oeller, P & Varma, H (2002) A draft sequence of the rice genome (*Oryza sativa* L. ssp. *japonica*). *Science,* 296, 92-100.
[http://dx.doi.org/10.1126/science.1068275]

Goffinet, B & Gerber, S (2000) Quantitative Trait Loci: A Meta-analysis. *Genetics,* 155, 463-73.

Göhre, V & Robatzek, S (2008) Breaking the barriers: microbial effector molecules subvert plant immunity. *Annu Rev Phytopathol,* 46, 189-215.
[http://dx.doi.org/10.1146/annurev.phyto.46.120407.110050]

Gómez-Gómez, L & Boller, T (2000) FLS2: an LRR receptor-like kinase involved in the perception of the bacterial elicitor flagellin in Arabidopsis. *Mol Cell,* 5, 1003-11.
[http://dx.doi.org/10.1016/s1097-2765(00)80265-8]

Gómez-Gómez, L, Bauer, Z & Boller, T (2001) Both the extracellular leucine-rich repeat domain and the kinase activity of FSL2 are required for flagellin binding and signaling in Arabidopsis. *Plant Cell,* 13, 1155-63.

Goudemand, E, Laurent, V, Duchalais, L, Ghaffary, SM, Kema, GH & Lonnet, P (2013) Association mapping and meta-analysis: two complementary approaches for the detection of reliable *Septoria tritici* blotch quantitative resistance in bread wheat (*Triticum aestivum* L.). *Mol Breed,* 32, 563-84.
[http://dx.doi.org/10.1007/s11032-013-9890-4]

Grandaubert, J, Balesdent, MH & Rouxel, T (2014) Evolutionary and adaptive role of transposable elements in fungal genomes.*Advances in Botanical Research,* Elsevier, Oxford, UK Vol. 70, 79-107.

Grant, SR, Fisher, EJ, Chang, JH, Mole, BM & Dangl, JL (2006) Subterfuge and manipulation: type III effector proteins of phytopathogenic bacteria. *Annu Rev Microbiol,* 60, 425-49.
[http://dx.doi.org/10.1146/annurev.micro.60.080805.142251]

Gray, YH (2000) It takes two transposons to tango: Transposable-element-mediated chromosomal rearrangements. *Trends Genet,* 16, 461-8.
[http://dx.doi.org/10.1016/s0168-9525(00)02104-1]

Greenberg, JT & Yao, N (2004) The role and regulation of programmed cell death in plant pathogen interactions. *Cell Microbiol,* 6, 201-11.
[http://dx.doi.org/10.1111/j.1462-5822.2004.00361.x]

Grube, RC, Radwanski, ER & Jahn, M (2000) Comparative genetics of disease resistance within the solanaceae. *Genetics,* 155, 873-87.

Grund, E, Tremousaygue, D & Deslandes, L (2019) Plant NLRs with Integrated Domains: Unity Makes Strength. *Plant Physiol,* 179, 1227-35.
[http://dx.doi.org/10.1104/pp.18.01134]

Guest, D & Brown, J (1980) *Plant defences against pathogens,* 263-86.

Guo, H & Ecker, JR (2004) The ethylene signaling pathway: new insights. *Curr Opin Plant Biol,* 7, 40-9.
[http://dx.doi.org/10.1016/j.pbi.2003.11.011]

Gupta, S, Mishra, VK, Kumari, S, Raavi Chand, R & Varadwaj, PK (2019) Deciphering genome-wide WRKY gene family of *Triticum aestivum* L. and their functional role in response to Abiotic stress. *Genes Genomics,* 41, 79-94.
[http://dx.doi.org/10.1007/s13258-018-0742-9]

Gururani, MA, Venkatesh, J, Upadhyaya, CP, Nookaraju, A, Pandey, SK & Park, SW (2012) Plant disease resistance genes: current status and future directions. *Physiol Mol Plant Pathol,* 78, 51-65.
[http://dx.doi.org/10.1016/j.pmpp.2012.01.002]

Gust, AA, Biswas, R, Lenz, HD, Rauhut, T, Ranf, S, Kemmerling, B, Götz, F, Glawischnig, E, Lee, J & Felix, G (2007) Bacteria-derived peptidoglycans constitute pathogen-associated molecular patterns triggering innate immunity in Arabidopsis. *J Biol Chem,* 282, 32338-48.
[http://dx.doi.org/10.1074/jbc.M704886200]

Haas, BJ, Kamoun, S, Zody, MC, Jiang, RH, Handsaker, RE, Cano, LM, Grabherr, M, Kodira, CD, Raffaele, S & Torto-Alalibo, T (2009) Genome sequence and analysis of the irish potato famine pathogen *Phytophthora infestans. Nature,* 461, 393.
[http://dx.doi.org/10.1038/nature08358]

Hain, R, Reif, HJ, Krause, E, Langebartels, R, Kindl, H, Vorna, B, Wiese, WE, Schmelzer, E, Schreier, PH,

Stöcker, RH & Strenzel, K (1993) Disease resistance results from foreign phytoalexin expression in a novel plant. *Nature,* 361, 153-6.
[http://dx.doi.org/10.1038/361153a0]

Hamilton, AJ & Baulcombe, DC (1999) A species of small antisense RNA in posttranscriptional gene silencing in plants. *Science,* 286, 950-2.
[http://dx.doi.org/10.1126/science.286.5441.950]

Hamilton, JP, Neeno-Eckwall, EC, Adhikari, BN, Perna, NT, Tisserat, N, Leach, JE, Lévesque, CA & Buell, CR (2011) The Comprehensive Phytopathogen Genomics Resource: a web-based resource for data-mining plant pathogen genomes. *Database (Oxford),* 2011, bar053.
[http://dx.doi.org/10.1093/database/bar053]

Hammond-Kosack, KE & Jones, JD (1997) Plant disease resistance genes. *Annu Rev Plant Physiol Plant Mol Biol,* 48, 575-607.
[http://dx.doi.org/10.1146/annurev.arplant.48.1.575]

Hamon, C, Baranger, A, Coyne, CJ, McGee, RJ, Le Goff, I & L'Anthoene, V (2011) New consistent QTL in pea associated with partial resistance to Aphanomyces euteiches in multiple French and American environments. *Theor Appl Genet,* 123, 261-81.
[http://dx.doi.org/10.1007/s00122-011-1582-z]

Han, F, Kleinhofs, A, Kilian, A & Ullrich, SE (1997) Clonning and mapping of a putative barley NADPH-dependent HC-toxine reductase. *Mol Plant Microbe Interact,* 10, 234-9.
[http://dx.doi.org/10.1094/MPMI.1997.10.2.234]

Hao, G, Pitino, M, Duan, Y & Stover, E (2016) Reduced Susceptibility to *Xanthomonas citri* in Transgenic Citrus Expressing the FLS2 Receptor From *Nicotiana benthamiana. Mol Plant Microbe Interact,* 29, 132-42.
[http://dx.doi.org/10.1094/MPMI-09-15-0211-R]

Hasegawa, M, Mitsuhara, I, Seo, S, Okada, K, Yamane, H, Iwai, T & Ohashi, Y (2014) Analysis on blast fungus-responsive characters of a flavonoid phytoalexin sakuranetin; Accumulation in infected rice leaves, antifungal activity and detoxification by fungus. *Molecules,* 19, 11404-18.
[http://dx.doi.org/10.3390/molecules190811404]

Hayashi, F, Smith, KD, Ozinsky, A, Hawn, TR, Yi, EC, Goodlett, DR, Eng, JK, Akira, S, Underhill, DM & Aderem, A (2001) The Innate Immune Response to Bacterial Flagellin Is Mediated by Toll-like Receptor 5. 410, 1099-3.

Hayashi, K & Yoshida, H (2009) Refunctionalization of the ancient rice blast disease resistance gene Pit by the recruitment of a retrotransposon as a promoter. *Plant J,* 57, 413-25.
[http://dx.doi.org/10.1111/j.1365-313X.2008.03694.x]

Hewezi, T, Lane, T, Piya, S, Rambani, A, Rice, JH & Staton, M (2017) Cyst nematode parasitism induces dynamic changes in the root epigenome. *Plant Physiol,* 174, 405.
[http://dx.doi.org/10.1104/pp.16.01948]

Hilder, VA, Gatehouse, AMR, Sheerman, SE, Barker, RF & Boulter, D (1987) A novel mechanism of insect resistance engineered into tobacco. *Nature,* 333, 160-3.
[http://dx.doi.org/10.1038/330160a0]

Hipskind, JD & Paiva, NL (2000) Constitutive accumulation of a resveratrol-glucoside in transgenic alfalfa increases resistance to *Phoma medicaginis. Mol Plant Microbe Interact,* 13, 551-62.
[http://dx.doi.org/10.1094/MPMI.2000.13.5.551]

Hirooka, T & Ishii, H (2013) Chemical control of plant diseases. *J Gen Plant Pathol,* 79, 390-401.
[http://dx.doi.org/10.1007/s10327-013-0470-6]

Holeva, MC, Bell, KS, Hyman, LJ, Avrova, AO, Whisson, SC, Birch, PRJ & Toth, IK (2004) Use of a pooled transposon mutation grid to demonstrate roles in disease development for *Erwinia carotovora* subsp. *atroseptica* putative type III secreted effector (DspE/A) and helper (HrpN) proteins. *Mol Plant Microbe Interact,* 17, 943-50.
[http://dx.doi.org/10.1094/mpmi.2004.17.9.943]

Horst, I, Welham, T, Kelly, S, Kaneko, T, Sato, S, Tabata, S, Parniske, M & Wang, TL (2007) TILLING mutants of *Lotus japonicus* reveal that nitrogen assimilation and fixation can occur in the absence of nodule-enhanced sucrose synthase. *Plant Physiol,* 144, 806-20.
[http://dx.doi.org/10.1104/pp.107.097063]

Hossard, L, Lannou, C, Papaix, J, Monod, H, Lô-Pezler, E, Souchère, V & Jeuffroy, MH (2010) Quel déploiement spatio-temporel des variétés et des itinéraires techniques pour accroître la durabilité des résistances variétales? *Innovations Agronomiques,* 8, 15-33.

Hovmøller, MS, Caffier, V, Jalli, M, Andersen, O & Besenhofer, G (2000) The European barley powdery mildew virulence survey and disease nursery 1993–1999. *Agronomie,* 20, 729-43.
[http://dx.doi.org/10.1051/agro:2000172]

Hu, X, Bidney, DL, Yalpani, N, Duvick, JP, Crasta, O, Folkerts, O & Lu, G (2003) Overexpression of a gene encoding hydrogen peroxide-generating oxalate oxidase evokes defense responses in sunflower. *Plant Physiol,* 133, 170-81.
[http://dx.doi.org/10.1104/pp.103.024026]

Hu, X, Xiao, G, Zheng, P, Shang, Y, Su, Y, Zhang, X, Liu, X, Zhan, S, St Leger, RJ & Wang, C (2014) Trajectory and genomic determinants of fungal-pathogen speciation and host adaptation. *Proc Natl Acad Sci USA,* 111, 16796-801.
[http://dx.doi.org/10.1073/pnas.1412662111]

Huang, L, Cheng, T, Xu, P, Fang, T & Xia, Q (2012) *Bombyx mori* transcription factors: genome-wide identification, expression profiles and response to pathogens by microarray analysis. *J Insect Sci,* 12, 40.
[http://dx.doi.org/10.1673/031.012.4001]

Huang, LK, Yan, HD, Zhao, XX, Zhang, XQ, Wang, J & Frazier, T (2015) Identifying differentially expressed genes under heat stress and developing molecular markers in orchardgrass (*Dactylis glomerata* L.) through transcriptome analysis. *Mol Ecol Resour,* 15, 1497-509.
[http://dx.doi.org/10.1111/1755-0998.12418]

Huang, X, Yan, HD, Zhang, XQ, Zhang, J, Frazier, TP & Huang, DJ (2016) *De novo* transcriptome analysis and molecular marker development of two Hemarthria species. *Front Plant Sci,* 7, 496.
[http://dx.doi.org/10.3389/fpls.2016.00496]

Huang, XH & Han, B (2014) Natural variations and genome-wide association studies in crop plants. *Annu Rev Plant Biol,* 65, 531-51.
[http://dx.doi.org/10.1146/annurev-arplant-050213-035715]

Huard-Chauveau, C, Perchepied, L, Debieu, M, Rivas, S, Kroj, T, Kars, I, Bergelson, J, Roux, F & Roby, D (2013) An atypical kinase under balancing selection confers broad-spectrum disease resistance in *Arabidopsis. PLoS Genet,* 9, e1003766.
[http://dx.doi.org/10.1371/journal.pgen.1003766]

Huibers, RP, Loonen, AE, Gao, D, Van den Ackerveken, G, Visser, RG & Bai, Y (2013) Powdery mildew resistance in tomato by impairment of SlPMR4 and SlDMR1. *PLoS One,* 8, e67467.
[http://dx.doi.org/10.1371/journal.pone.0067467]

Hulbert, SH, Webb, CA, Smith, SM & Sun, Q (2001) Resistance gene complexes: evolution and utilization. *Annu Rev Phytopathol,* 39, 285-312.
[http://dx.doi.org/10.1146/annurev.phyto.39.1.285]

Hwang, CF, Bhakta, AV, Truesdell, GM, Pudlo, WM & Moroz Williamson, V (2000) Evidence for a role of the N terminus and leucine rich repeats region of the *Mi* gene product in regulation of localized cell death. *Plant Cell,* 12, 1319-30.
[http://dx.doi.org/10.1105/tpc.12.8.1319]

Ikeda, K, Kitagawa, H, Shimoi, S, Inoue, K, Osaki, Y & Nakayashiki, H (2013) Attachment of airborne pathogens to their host: a potential target for disease control. *Acta Phytopathologica Sin,* 43, 18.

Innes, RW (2004) Guarding the goods. New insights into the central alarm system of plants. *Plant Physiol,* 135, 695-701.
[http://dx.doi.org/10.1104/pp.104.040410]

Inohara, N & Nunez, G (2001) The NOD: a signaling module that regulates apoptosis and host defense against pathogens. *Oncogene,* 20, 6473-81.
[http://dx.doi.org/10.1038/sj.onc.1204787]

Ishibashi, K, Kezuka, Y, Kobayashi, C, Kato, M, Inoue, T, Nonaka, T, Ishikawa, M, Matsumura, H & Katoh, E (2014) Structural basis for the recognition-evasion arms race between Tomato mosaic virus and the resistance gene Tm-1. *Proc Natl Acad Sci USA,* 111, E3486-95.
[http://dx.doi.org/10.1073/pnas.1407888111]

Jabs, T, Tschöpe, M, Colling, C, Hahlbrock, K & Scheel, D (1997) Elicitor-stimulated ion fluxes and O2− from the oxidative burst are essential components in triggering defense gene activation and phytoalexin synthesis in parsley. *Proc Natl Acad Sci USA,* 94, 4800-5.
[http://dx.doi.org/10.1073/pnas.94.9.4800]

Jackson Seukep, A, Noumedem, JAK, Djeussi, DE & Kuete, V (2014) Genotoxicity and Teratogenicity of African Medicinal Plants.*Toxicological Survey of African Medicinal Plants,* Elsevier Inc. 235-75.

James, C (2015) *20th Anniversary (1996 to 2015) of the Global Commercialization of Biotech Crops and Biotech Crop Highlights in 2015*ISAAA Brief No. 51. ISAAA: Ithaca, NY. 978-1-892456-65-6.www.isaaa.org

Jamieson, PA, Shan, L & He, P (2018) Plant cell surface molecular cypher: Receptor-like proteins and their roles in immunity and development. *Plant Sci,* 274, 242-51.
[http://dx.doi.org/10.1016/j.plantsci.2018.05.030]

Janeway, CA, Jr & Medzhitov, R (2002) Innate Immune Recognition. *Annu Rev Immunol,* 20, 197-216.
[http://dx.doi.org/10.1146/annurev.immunol.20.083001.084359]

Jangam, D, Feschotte, C & Betrán, E (2017) Transposable Element Domestication As an Adaptation to Evolutionary Conflicts. *Trends Genet,* 33, 817-31.
[http://dx.doi.org/10.1016/j.tig.2017.07.011]

Jankowicz-Cieslak, J, Mba, C & Till, BJ (2017) Mutagenesis for Crop Breeding and Functional Genomics.*Biotechnologies for Plant Mutation Breeding: Protocols,* Springer International Publishing 3-18.

Jeandet, P, Bessis, R, Sbaghi, M & Meunier, P (1995) Production of the Phytoalexin Resveratrol by Grapes as a Response to Botrytis Attack Under Natural Conditions. *J Phytopathol,* 143, 135-9.
[http://dx.doi.org/10.1111/j.1439-0434.1995.tb00246.x]

Jeong, JJ, Ju, HJ & Noh, J (2014) A Review of Detection Methods for the Plant Viruses. *Singmulbyeong Yeon-gu,* 20, 173-81.
[http://dx.doi.org/10.5423/RPD.2014.20.3.173]

Jha, G, Thakur, K & Thakur, P (2009) The *Venturia* apple pathosystem: pathogenicity mechanisms and plant defense responses. *J Biomed Biotechnol,* 680160
[http://dx.doi.org/10.1155/2009/680160]

Jha, S & Chattoo, BB (2010) Expression of a plant defensin in rice confers resistance to fungal phytopathogens. *Transgenic Res,* 19, 373-84.
[http://dx.doi.org/10.1007/s11248-009-9315-7]

Jia, J, Zhao, S & Kong, X (2013) *Aegilops tauschii* draft genome sequence reveals a gene repertoire for wheat adaptation. *Nature,* 496, 91-5.
[http://dx.doi.org/10.1038/nature12028]

Jia, Y, McAdams, SA, Bryan, GT, Hershey, HP & Valent, B (2000) Direct interaction of resistance gene and avirulence gene products confers rice blast resistance. *EMBO J,* 19, 4004-14.
[http://dx.doi.org/10.1093/emboj/19.15.4004]

Jiang, RHY & Tyler, BM (2012) Mechanisms and Evolution of Virulence in Oomycetes. *Annu Rev Phytopathol,* 50, 295-318.
[http://dx.doi.org/10.1146/annurev-phyto-081211-172912]

Jibril, SM, Jakada, BH, Kutama, AS & Umar, HY (2016) Plant and Pathogens: Pathogen Recognision, Invasion and Plant Defense Mechanism. *Int J Curr Microbiol Appl Sci,* 5, 247-57.
[http://dx.doi.org/10.20546/ijcmas.2016.506.028]

Xu, J & Wang, N (2019) Where are we going with genomics in plant pathogenic bacteria? *Genomics,* 111, 729-36.
[http://dx.doi.org/10.1016/j.ygeno.2018.04.011]

Johal, G & Briggs, S (1992) Reductase activity encoded by the HM1 disease resistance gene in maize. *Science,* 258, 985-7.
[http://dx.doi.org/10.1126/science.1359642]

Johnson, C, Boden, E & Arias, J (2003) Salicylic acid and NPR1 induce the recruitment of trans-activating TGA factors to a defense gene promoter in Arabidopsis. *Plant Cell,* 15, 1846-58.
[http://dx.doi.org/10.1105/tpc.012211]

Johnson, KP, Malenke, JR & Clayton, DH (2009) Competition promotes the evolution of host generalists in obligate parasites. *Proc R Soc Lond B Biol Sci,* 276, 3921-6.
[http://dx.doi.org/10.1098/rspb.2009.1174]

Jonak, C, Heberle-Bors, E & Hirt, H (1994) MAP kinases: universal multi-purpose signaling tools. *Plant Mol Biol,* 24, 407-16.
[http://dx.doi.org/10.1007/BF00024109]

Jones, DA & Jones, JDG (1997) The role of leucine-rich repeat proteins in plant defences. *Adv Bot Res,* 24, 90-167.

Jones, DA, Thomas, CM, Hammond-Kosack, KE, Balint-Kurti, PJ & Jones, JD (1994) Isolation of the tomato *Cf-9* gene for resistance to *Cladosporium fulvum* by transposon tagging. *Science,* 266, 789-93.
[http://dx.doi.org/10.1126/science.7973631]

Jones, JD & Dangl, JL (2006) The plant immune system. *Nature,* 444, 323-9.
[http://dx.doi.org/10.1038/nature05286]

Jones, JD, Thomas, CM, Hammond-Kosack, KE, Balint-Kurti, PJ & Jones, DA (1993) Two complex resistance loci revealed in tomato by classical and RFLP mapping of the *Cf-2, Cf-4, Cf-5* and *Cf-9* genes for resistance to Cladosporium fulvum. *Mol Plant Microbe Interact,* 6, 348-57.
[http://dx.doi.org/10.1094/MPMI-6-348]

Jordan, IK, Rogozin, IB, Glazko, GV & Koonin, EV (2003) Origin of a substantial fraction of human regulatory sequences from transposable elements. *Trends Genet,* 19, 68-72.
[http://dx.doi.org/10.1016/s0168-9525(02)00006-9]

Jorgensen, JH (1993) Durability of resistance in the pathosystem barley-powedry mildew. In: Jacobs, T.H., Parlevliet, J.E., (Eds.), *Durability of disease resistance,* Springer Science, Business Media Dordrecht.

José-Estanyol, M, Gomis-Rüth, FX & Puigdomènech, P (2004) The eight-cysteine motif, a versatile structure in plant proteins. *Plant Physiol Biochem,* 42, 355-65.
[http://dx.doi.org/10.1016/j.plaphy.2004.03.009]

Joshi, RK & Nayak, S (2011) Functional characterization and signal transduction ability of nucleotide-binding site-leucine-rich repeat resistance genes in plants. *Genet Mol Res,* 10, 2637-52.
[http://dx.doi.org/10.4238/2011.October.25.10]

Jourdan, E, Ongena, M & Thonart, P (2008) Caractéristiques moléculaires de l'immunité des plantes induite par les rhizobactéries non pathogènes. *Biotechnol Agron Soc Environ,* 12, 437-49.

Jubic, LM, Saile, S, Furzer, OJ, El Kasmi, F & Dangl, JL (2019) Help wanted: helper NLRs and plant immune responses. *Curr Opin Plant Biol,* 50, 82-94.
[http://dx.doi.org/10.1016/j.pbi.2019.03.013]

Kaku, H, Nishizawa, Y, Ishii-Minami, N, Akimoto-Tomiyama, C, Dohmae, N, Takio, K, Minami, E & Shibuya, N (2006) Plant cells recognize chitin fragments for defense signaling through a plasma membrane receptor. *Proc Natl Acad Sci USA,* 103, 11086-91.
[http://dx.doi.org/10.1073/pnas.0508882103]

Kaloshian, I, Lange, WH & Williamson, VM (1995) An aphid-resistance locus is tightly linked to the nematode-resistance gene, *Mi,* in tomato. *Proc Natl Acad Sci USA,* 92, 622-5.
[http://dx.doi.org/10.1073/pnas.92.2.622]

Kamoun, S, Furzer, O, Jones, JDG, Judelson, HS & Ali, GS (2014) The Top 10 oomycete pathogens in molecular plant pathology. *Mol Plant Pathol,* 16, 413-34.
[http://dx.doi.org/10.1111/mpp.12190]

Kamoun, S (2006) A catalogue of the effector secretome of plant pathogenic oomycetes. *Annu Rev Phytopathol,* 44, 41-60.
[http://dx.doi.org/10.1146/annurev.phyto.44.070505.143436]

Kang, S, Lebrun, MH, Farrall, L & Valent, B (2001) Gain of virulence caused by insertion of a Pot3 transposon in a *Magnaporthe grisea* avirulence gene. *Mol Plant Microbe Interact,* 14, 671-4.

[http://dx.doi.org/10.1094/MPMI.2001.14.5.671]

Kang, YJ, Kim, KH, Shim, S, Yoon, MY, Sun, S, Kim, MY, Van, K & Lee, S-H (2012) Genome-wide mapping of NBS-LRR genes and their association with disease resistance in soybean. *BMC Plant Biol,* 12, 139.
[http://dx.doi.org/10.1186/1471-2229-12-139]

Kanzaki, H, Yoshida, K, Saitoh, H, Fujisaki, K, Hirabuchi, A, Alaux, L, Fournier, E, Tharreau, D & Terauchi, R (2012) Arms race co-evolution of *Magnaporthe oryzae AVR-Pik* and rice *Pik* genes driven by their physical interactions. *Plant J,* 72, 894-907.
[http://dx.doi.org/10.1111/j.1365-313X.2012.05110.x]

Katagiri, F (2004) A global view of defense gene expression regulation-a highly interconnected signaling network. *Curr Opin Plant Biol,* 7, 506-11.
[http://dx.doi.org/10.1016/j.pbi.2004.07.013]

Kauffmann, S, Dorey, S & Fritig, B (1999) Les stratégies de défense des plantes. *Pour la Science,* 262, 30-7.

Kawakatsu, T, Huang, SC, Jupe, F, Sasaki, E, Schmitz, RJ, Urich, MA, Castanon, R, Nery, JR, Barragan, C & He, Y (2016) Epigenomic Diversity in a Global Collection of Arabidopsis thaliana Accessions. *Cell,* 166, 492-505.
[http://dx.doi.org/10.1016/j.cell.2016.06.044]

Keen, NT (1975) Specific elicitors of plant phytoalexin production: determinants of race specificity in pathogens? *Science,* 187, 74-5.
[http://dx.doi.org/10.1126/science.187.4171.74]

Kemmerling, B, Halter, T, Mazzotta, S, Mosher, S & Nürnberger, T (2011) A genome-wide survey for Arabidopsis leucine-rich repeat receptor kinases implicated in plant immunity. *Front Plant Sci,* 2, 88.
[http://dx.doi.org/10.3389/fpls.2011.00088]

Kesarwani, M, Yoo, J & Dong, X (2007) Genetic interactions of TGA transcription factors in the regulation of pathogenesis-related genes and disease resistance in Arabidopsis. *Plant Physiol,* 144, 336-46.
[http://dx.doi.org/10.1104/pp.106.095299]

Kessler, A & Balwin, IT (2002) Plant responses to insect herbivory: the emerging molecular analysis. *Annu Rev Plant Biol,* 53, 299-328.
[http://dx.doi.org/10.1146/annurev.arplant.53.100301.135207]

Key, S, Ma, JK & Drake, PM (2008) Genetically modified plants and human health. *J R Soc Med,* 101, 290-8.
[http://dx.doi.org/10.1258/jrsm.2008.070372]

Khatib, M, Lafitte, C, Esquerré-Tugayé, M-T, Bottin, A & Rickauer, M (2004) The CBEL elicitor of *Phytophthora parasitica var. nicotianae* activates defence in *Arabidopsis thalianavia* three different signalling pathways. *New Phytol,* 162, 501-10.
[http://dx.doi.org/10.1111/j.1469-8137.2004.01043.x]

Khedia, J, Agarwal, P & Agarwal, PK (2018) AlNAC4 Transcription Factor From Halophyte Aeluropus lagopoides Mitigates Oxidative Stress by Maintaining ROS Homeostasis in Transgenic Tobacco. *Front Plant Sci,* 9, 1522.
[http://dx.doi.org/10.3389/fpls.2018.01522]

Khong, GN, Richaud, F, Coudert, Y, Pati, PK & Santi, C (2008) Modulating rice stress tolerance by transcription factors. *Biotechnol Genet Eng Rev,* 25, 381-403.

[http://dx.doi.org/10.5661/bger-25-381]

Kim, JF & Beer, SV (2000) Hrp genes and harpins of Erwinia amylovora: A decade of discovery.*Fire blight: the disease and its causative agent, Erwinia amylovora,* CABI, Wallingford, UK 141-61.

King, A & Young, G (1999) Characteristics and Occurrence of Phenolic Phytochemicals. *J Am Diet Assoc,* 99, 213-8.
[http://dx.doi.org/10.1016/s0002-8223(99)00051-6]

Kissoudis, C, Chowdhury, R, van Heusden, S, van de Wiel, C, Finkers, R, Visser, RGF, Bai, Y & van der Linden, G (2015) Combined biotic and abiotic stress resistance in tomato. *Euphytica,* 202, 317-32.
[http://dx.doi.org/10.1007/s10681-015-1363-x]

Kitajima, S & Sato, F (1999) Plant Pathogenesis-Related Proteins: Molecular Mechanisms of Gene Expression and Protein Function. *J Biochem,* 125, 1-8.
[http://dx.doi.org/10.1093/oxfordjournals.jbchem.a022244]

Kiyosawa, S (1982) Genetics and epidemiological modeling of breakdown of plant-disease resistance. *Annu Rev Phytopathol,* 20, 93-117.

Klarzynski, O & Fritig, B (2001) Stimulation des défenses naturelles des plantes. *Life Sci,* 324, 953-63.
[http://dx.doi.org/10.1016/s0764-4469(01)01371-3]

Kloepper, JW (1991) Plant growth-promoting rhizobacteria as biological control agents of soilborne diseases.*Biological control of plant diseases,* Food and Fertilizer Technology Center, Taiwan 142-56.

Kloepper, JW, Ryu, CM & Zhang, S (2004) Induced Systemic Resistance and Promotion of Plant Growth by Bacillus spp. *Phytopathology,* 94, 1259-66.
[http://dx.doi.org/10.1094/phyto.2004.94.11.1259]

Kloepper, JW & Schroteh, MN (1981) Plant growth-promoting rhizobacteria and plant growth under gnotobiotic conditions. *Phytopathology,* 71, 642-4.
[http://dx.doi.org/10.1094/Phyto-71-642]

Knights, EJ & Hobson, KB (2016) Chickpea overview. *Reference Module in Food Science,* Elsevier.
[http://dx.doi.org/10.1016/B978-0-08-100596-5.00035-4]

Kobe, B & Diessenhofer, J (1995) A structural basis of the interactions between leucine rich repeats and protein ligands. *Nature,* 374, 183-6.
[http://dx.doi.org/10.1038/374183a0]

Kobe, B & Diessenhofer, J (1995) Proteins with leucine-rich repeats. *Curr Opin Struct Biol,* 5, 409-16.
[http://dx.doi.org/10.1016/0959-440x(95)80105-7]

Kobe, B & Kajava, AV (2000) When protein folding is simplified to protein coiling: the continuum of solenoid protein structures. *Trends Biochem Sci,* 25, 509-15.
[http://dx.doi.org/10.1016/s0968-0004(00)01667-4]

Kobe, B & Kajava, AV (2001) The leucine-rich repeat as a protein recognition motif. *Curr Opin Struct Biol,* 11, 725-32.
[http://dx.doi.org/10.1016/s0959-440x(01)00266-4]

Kohler, A, Rinaldi, C, Duplessis, S, Baucher, M, Geelen, D, Duchaussoy, F, Meyers, BC, Boerjan, W & Martin, F (2008) Genome-wide identification of NBS resistance genes in *Populus trichocarpa. Plant Mol Biol,* 66, 619-36.

[http://dx.doi.org/10.1007/s11103-008-9293-9]

Kolmer, JA & Leonard, KJ (1986) Genetic selection and adaptation of Cochliobolus heterostrophus to corn host with partial resistance. *Phytopathology,* 76, 774-7.
[http://dx.doi.org/10.1094/Phyto-76-774]

Kombrink, E & Schmelzer, E (2001) The Hypersensitive Response and its Role in Local and Systemic Disease Resistance. *Eur J Plant Pathol,* 107, 69-78.
[http://dx.doi.org/10.1023/A:1008736629717]

Kombrink, E & Somssich, IE (1997) Pathogenesis-related proteins and plant defense.*The Mycota V, Part A: Plant relationships,* Springer-Verlag, Berlin 107-28.

Kou, YJ & Wang, SP (2010) Broad-spectrum and durability: understanding of quantitative disease resistance. *Curr Opin Plant Biol,* 13, 181-5.
[http://dx.doi.org/10.1016/j.pbi.2009.12.010]

Koul, B, Srivastava, S, Sanyal, I, Tripathi, B, Sharma, V & Amla, DV (2014) Transgenic tomato line expressing modified Bacillus thuringiensis cry1Ab gene showing complete resistance to two lepidopteran pests. *Springerplus,* 3, 84.
[http://dx.doi.org/10.1186/2193-1801-3-84]

Kourelis, J & van der Hoorn, RAL (2018) Defended to the Nines: 25 years of resistance gene cloning identifies nine mechanisms for R protein function. *Plant Cell,* 30, 285-99.
[http://dx.doi.org/10.1105/tpc.17.00579]

Kover, PX & Caicedo, AL (2001) The genetic architecture of disease resistance in plants and the maintenance of recombination by parasites. *Mol Ecol,* 10, 1-16.
[http://dx.doi.org/10.1046/j.1365-294X.2001.01124.x]

Krasileva, KV, Dahlbeck, D & Staskawicz, BJ (2010) Activation of an Arabidopsis Resistance Protein Is Specified by the in Planta Association of Its Leucine-Rich Repeat Domain with the Cognate Oomycete Effector. *Plant Cell,* 22, 2444-58.
[http://dx.doi.org/10.1105/tpc.110.075358]

Krattinger, SG & Keller, B (2016) Molecular genetics and evolution of disease resistance in cereals. *New Phytol,* 212, 320-32.
[http://dx.doi.org/10.1111/nph.14097]

Kroj, T, Chanclud, E, Michel-Romiti, C, Grand, X & Morel, JB (2016) Integration of decoy domains derived from protein targets of pathogen effectors into plant immune receptors is widespread. *New Phytol,* 210, 618-26.
[http://dx.doi.org/10.1111/nph.13869]

Kundu, A & Vadassery, J (2019) Chlorogenic acid-mediated chemical defence of plants against insect herbivores. Plant Biology 21, 185-9.
[http://dx.doi.org/10.1111/plb.12947]

Kuhn, T (1962) *The Structure of Scientific Revolutions,* University of Chicago Press, Chicago.

Kunze, G, Zipfel, C, Robatzek, S, Niehaus, K, Boller, T & Felix, G (2004) The N terminus of bacterial elongation factor Tu elicits innate immunity in Arabidopsis plants. *Plant Cell,* 16, 3496-507.
[http://dx.doi.org/10.1105/tpc.104.026765]

Kurowska, M, Daszkowska-Golec, A & Gruszka, D (2011) TILLING: a shortcut in functional genomics. *J*

Appl Genet, 52, 371-90.
[http://dx.doi.org/10.1007/s13353-011-0061-1]

Kusch, S & Panstruga, R (2017) mlo-Based Resistance: An Apparently Universal Weapon to Defeat Powdery Mildew Disease. *Mol Plant Microbe Interact,* 30, 179-89.
[http://dx.doi.org/10.1094/MPMI-12-16-0255-CR]

Kushalappa, AC, Yogendra, KN & Karre, S (2016) Plant innate immune response: qualitative and quantitative resistance. *Crit Rev Plant Sci,* 35, 38-55.
[http://dx.doi.org/10.1080/07352689.2016.1148980]

Lacombe, S, Rougon-Cardoso, A, Sherwood, E, Peeters, N, Dahlbeck, D, van Esse, HP, Smoker, M, Rallapalli, G, Thomma, BP, Staskawicz, B, Jones, JD & Zipfel, C (2010) Interfamily transfer of a plant pattern-recognition receptor confers broad-spectrum bacterial resistance. *Nat Biotechnol,* 28, 365-9.
[http://dx.doi.org/10.1038/nbt.1613]

Lake, JA, Field, KJ, Davey, MP, Beerling, DJ & Lomax, BH (2009) Metabolomic and physiological responses reveal multi-phasic acclimation of *Arabidopsis thaliana* to chronic UV radiation. *Plant Cell Environ,* 32, 1377-89.
[http://dx.doi.org/10.1111/j.1365-3040.2009.02005.x]

Lanfermeijer, FC, Warmink, J & Hille, J (2005) The Products of the Broken Tm-2 and the Durable Tm-2(2) Resistance Genes From Tomato Differ in Four Amino Acids. *J Exp Bot,* 56, 2925-33.
[http://dx.doi.org/10.1093/jxb/eri288]

Lannoo, N & Van Damme, EJ (2014) Lectin domains at the frontiers of plant defense. *Front Plant Sci,* 5, 397.
[http://dx.doi.org/10.3389/fpls.2014.00397]

Lapin, D, Kovacova, V, Sun, X, Dongus, JA, Bhandari, DD, von Born, P, Bautor, J, Guarneri, N, Rzemieniewski, J, Stuttmann, J, Beyer, A & Parkera, JE (2019) A coevolved EDS1-SAG101-NRG1 module mediates cell death signaling by TIR-domain immune receptors. *The Plant Cell,* 31, 2430-55.
[http://dx.doi.org/10.1105/tpc.19.00118]

Lavaud, C (2015) *Diversité et combinaison des modes d'actions des QTL de résistance à Aphanomyces euteiches chez le pois*PhD Thesis, Agrocampus Ouest.https://tel.archives-ouvertes.fr/tel-01493826/document [In French]

Lawrence, GJ, Finnegan, EJ, Ayliffe, MA & Ellis, JG (1995) The L6 gene for flax rust resistance is related to the Arabidopsis bacterial resistance gene RPS2 and the tobacco viral resistance gene N. *Plant Cell,* 7, 1195-206.
[http://dx.doi.org/10.1105/tpc.7.8.1195]

Le, TN, Schumann, U, Smith, NA, Tiwari, S, Au, PCK, Zhu, QH, Taylor, JM, Kazan, K, Llewellyn, DJ & Zhang, R (2014) DNA demethylases target promoter transposable elements to positively regulate stress responsive genes in Arabidopsis. *Genome Biol,* 15, 458.
[http://dx.doi.org/10.1186/s13059-014-0458-3]

Lee, H, Cha, J, Choi, C, Choi, N, Ji, J, Park, S, Lee, S & Hwang, D (2018) Rice WRKY11 Plays a Role in Pathogen Defense and Drought Tolerance. *Rice,* 11, 5.
[http://dx.doi.org/10.1186/s12284-018-0199-0]

Lefebvre, V & Palloix, A (1996) Both epistatic and additive effects of QTLs are involved in polygenic induced resistance to disease: a case study, the interaction pepper - Phytophthora capsici Leonian. *Theor Appl Genet,* 93, 503-11.

[http://dx.doi.org/10.1007/bf00417941]

Lehman, JS & Shaner, G (1997) Selection of populations of Puccinia recondita f. sp. tritici for shortened latent period on a partially resistant wheat cultivar. *Phytopathology,* 87, 170-6.
[http://dx.doi.org/10.1094/phyto.1997.87.2.170]

Leitao, JM (2011) Plant mutation breeding and biotechnology. In: Shu, Q.Y., Forster, B.P., Nakagawa, H., (Eds.), *Chemical mutagenesis,* CABI, Wallingford 135-58.

Lemarie, S, Robert-Seilaniantz, A, Lariagon, C, Lemoine, J, Marnet, N & Levrel, A (2015) Camalexin contributes to the partial resistance of *Arabidopsis thaliana* to the biotrophic soilborne protist *Plasmodiophora brassicae. Front Plant Sci,* 6, 539.
[http://dx.doi.org/10.3389/fpls.2015.00539]

Lewis, JD, Lee, AH-Y, Hassan, JA, Wan, J, Hurley, B, Jhingree, JR, Wang, PW, Lo, T, Youn, J-Y, Guttman, DS & Desveaux, D (2013) The *Arabidopsis* ZED1 pseudokinase is required for ZAR1-mediated immunity induced by the *Pseudomonas syringae* type III effector HopZ1a. *Proc Natl Acad Sci USA,* 110, 18722-7.
[http://dx.doi.org/10.1073/pnas.1315520110]

Lewis, JD, Wu, R, Guttman, DS & Desveaux, D (2010) Allele-Specific Virulence Attenuation of the *Pseudomonas syringae* HopZ1a Type III Effector *via* the Arabidopsis ZAR1 Resistance Protein. *PLoS Genet,* 6, e1000894.
[http://dx.doi.org/10.1371/journal.pgen.1000894]

Lewsey, MG, Hardcastle, TJ, Melnyk, CW, Molnar, A, Valli, A, Urich, MA, Nery, JR, Baulcombe, DC & Ecker, JR (2016) Mobile small RNAs regulate genome-wide DNA methylation. *Proc Natl Acad Sci USA,* 113, E801-10.
[http://dx.doi.org/10.1073/pnas.1515072113]

Li, C, Wang, D & Peng, S (2019) Genome-wide association mapping of resistance against rice blast strains in South China and identification of a new *Pik* allele. *Rice (N Y),* 12, 47.
[http://dx.doi.org/10.1186/s12284-019-0309-7]

Li, F & Ding, SW (2006) Virus counterdefense: diverse strategies for evading the RNA-silencing immunity. *Annu Rev Microbiol,* 60, 503-31.
[http://dx.doi.org/10.1146/annurev.micro.60.080805.142205]

Li, J, Wang, L, Zhan, Q, Liu, Y & Yang, X (2016) Transcriptome characterization and functional marker development in *Sorghum sudanense. PLoS One,* 11, e0154947.
[http://dx.doi.org/10.1371/journal.pone.0154947]

Li, Y, Zhong, Y, Huang, K & Cheng, Z-M (2016) Genome wide analysis of NBS-encoding genes in kiwi fruit (*Actinidia chinensis*). *J Genet,* 95, 997-1001.
[http://dx.doi.org/10.1007/s12041-016-0700-8]

Liao, J, Huang, H, Meusnier, I, Adreit, H, Ducasse, A & Bonnot, F (2016) Pathogen effectors and plant immunity determine specialization of the blast fungus to rice subspecies. *eLife,* 5, e19377.
[http://dx.doi.org/10.7554/eLife.19377]

Liégard, B (2018) Rôles des variations épigénétiques transgénérationnelles dans la résistance quantitative à la hernie chez Arabidopsis thaliana. *PhD Thesis,*https://tel.archives-ouvertes.fr/tel-02303309/ In French

Liégard, B, Baillet, V, Etcheverry, M, Joseph, E & Lariagon, C (2019) Quantitative resistance to clubroot infection mediated by transgenerational epigenetic variation in Arabidopsis. *New Phytol,* 222, 468-79.

[http://dx.doi.org/10.1111/nph.15579]

Ligterink, W & Hirt, H (2001) Mitogen-activated protein [MAP] kinase pathways in plants, versatile signaling tools. *Int Rev Cytol,* 201, 209-75.
[http://dx.doi.org/10.1016/S0074-7696(01)01004-X]

Lindhout, P (2002) The perspectives of polygenic resistance in breeding for durable disease resistance. *Euphytica,* 124, 217-26.
[http://dx.doi.org/10.1023/A:1015686601404]

Ling, H, Zhao, S & Liu, D (2013) The perspectives of polygenic resistance in breeding for durable disease resistance. *Euphytica,* 124, 217-26.

Lister, R, O'Malley, RC, Tonti-Filippini, J, Gregory, BD, Berry, CC, Millar, AH & Ecker, JR (2008) Highly integrated single-base resolution maps of the epigenome in Arabidopsis. *Cell,* 133, 523-36.
[http://dx.doi.org/10.1016/j.cell.2008.03.029]

Liu, Q, Yuan, M, Zhou, Y, Li, X, Xiao, J & Wang, S (2011) A paralog of the MtN3/saliva family recessively confers race-specific resistance to Xanthomonas oryzae in rice. *Plant Cell Environ,* 34, 1958-69.
[http://dx.doi.org/10.1111/j.1365-3040.2011.02391.x]

Liu, R, Lü, B, Wang, X, Zhang, C, Zhang, S, Qian, J, Chen, L, Shi, H & Dong, H (2010) Thirty-seven transcription factor genes differentially respond to a hairpin protein and affect resistance to the green peach aphid in *Arabidopsis. J Biosci,* 35, 435-50.
[http://dx.doi.org/10.1007/s12038-010-0049-8]

Liu, Z, Wu, Y, Yang, F, Zhang, Y, Chen, S, Xie, Q, Tian, X & Zhou, J-M (2013) BIK1 interacts with PEPRs to mediate ethylene-induced immunity. *Proc Natl Acad Sci USA,* 110, 6205-10.
[http://dx.doi.org/10.1073/pnas.1215543110]

Llorens, E, García-Agustín, P & Lapeña, L (2017) Advances in induced resistance by natural compounds: towards new options for woody crop protection. *Sci Agric,* 74, 90-100.
[http://dx.doi.org/10.1590/1678-992x-2016-0012]

Loake, G & Grant, M (2007) Salicylic acid in plant defence - the players and protagonists. *Curr Opin Plant Biol,* 10, 466-72.
[http://dx.doi.org/10.1016/j.pbi.2007.08.008]

Lopes, MA, Hora, BT, Dias, CV, Santos, GC, Gramacho, KP, Cascardo, JC, Gesteira, AS & Micheli, F (2010) Expression analysis of transcription factors from the interaction between cacao and *Moniliophthora Perniciosa* (*Tricholomataceae*). *Genet Mol Res,* 9, 1279-97.
[http://dx.doi.org/10.4238/vol9-3gmr825]

López Sánchez, A, Stassen, JH, Furci, L, Smith, LM & Ton, J (2016) The role of DNA (de)methylation in immune responsiveness of Arabidopsis. *Plant J,* 88, 361-74.
[http://dx.doi.org/10.1111/tpj.13252]

Lorang, J, Kidarsa, T, Bradford, CS, Gilbert, B, Curtis, M, Tzeng, SC, Maier, CS & Wolpert, TJ (2012) Tricking the guard: exploiting plant defense for disease susceptibility. *Science,* 338, 659-62.
[http://dx.doi.org/10.1126/science.1226743]

Lorenzo, O & Solano, R (2005) Molecular players regulating the jasmonate signalling network. *Curr Opin Plant Biol,* 8, 532-40.
[http://dx.doi.org/10.1016/j.pbi.2005.07.003]

Lotze, MT, Deisseroth, A & Rubartelli, A (2007) Damage associated molecular pattern molecules. *Clin Immunol,* 124, 1-4.

Lozano, R, Ponce, O, Ramirez, M, Mostajo, N & Orjeda, G (2012) Genome-Wide Identification and Mapping of NBS-Encoding Resistance Genes in *Solanum tuberosum* Group Phureja. *PLoS One,* 7, e34775. [http://dx.doi.org/10.1371/journal.pone.0034775]

Lozano-Torres, JL, Wilbers, RHP, Gawronski, P, Boshoven, JC, Finkers-Tomczak, A, Cordewener, JHG, America, AHP, Overmars, HA, van't Klooster, JW, Baranowski, L, Sobczak, M, Ilyas, M, van der Hoorn, RAL & Schots, A (2012) Dual disease resistance mediated by the immune receptor Cf-2 in tomato requires a common virulence target of a fungus and a nematode. *Proceedings of the National Academy of Sciences USA* 109, 10119-24.

Lu, D, Wu, S, Gao, X, Zhang, Y, Shan, L & He, P (2010) A receptor-like cytoplasmic kinase, BIK1, associates with a flagellin receptor complex to initiate plant innate immunity. 107, 496-501.

Lu, F, Wang, H, Wang, S, Jiang, W, Shan, C, Li, B, Yang, J, Zhang, S & Sun, W (2015) Enhancement of innate immune system in monocot rice by transferring the dicotyledonous elonga- tion factor Tu receptor EFR. *J Integr Plant Biol,* 57, 641-52. [http://dx.doi.org/10.1111/jipb.12306]

Luck, JE, Lawrence, GJ, Dodds, PN, Shepherd, KW & Ellis, JG (2000) Regions outside of the leucine-rich repeats of flax rust resistance proteins play a role in specificity determination. *Plant Cell,* 12, 1367-77. [http://dx.doi.org/10.1105/tpc.12.8.1367]

Luderer, R, Takken, FLW, de Wit, PJGM & Joosten, MHAJ (2002) Cladosporium fulvum overcomes Cf-- -mediated resistance by producing truncated AVR2 elicitor proteins. *Mol Microbiol,* 45, 875-84. [http://dx.doi.org/10.1046/j.1365-2958.2002.03060.x]

Luna, E, Bruce, TJ, Roberts, MR, Flors, V & Ton, J (2012) Next-generation systemic acquired resistance. *Plant Physiol,* 158, 844-53. [http://dx.doi.org/10.1104/pp.111.187468]

Lyon, GD, Reglinski, T & Newton, AC (1995) Novel disease control compounds: the potential to immunize plants against infection. *Plant Pathol,* 44, 407-27. [http://dx.doi.org/10.1111/j.1365-3059.1995.tb01664.x]

Macho, AP & Zipfel, C (2014) Plant PRRs and the Activation of Innate Immune Signaling. *Mol Cell,* 54, 263-72. [http://dx.doi.org/10.1016/j.molcel.2014.03.028]

Mackey, D, Belkhadir, Y, Alonso, JM, Ecker, JR & Dangl, JL (2003) *Arabidopsis* RIN4 Is a Target of the Type III Virulence Effector AvrRpt2 and Modulates RPS2-Mediated Resistance. *Cell,* 112, 379-89. [http://dx.doi.org/10.1016/s0092-8674(03)00040-0]

Mackey, D, Belkhadir, Y, Alonso, JM, Ecker, JR & Dangl, JL (2003) Arabidopsis RIN4 is a target of the type III virulence effector AvrRpt2 and modulates RPS2-mediated resistance. *Cell,* 112, 379-89. [http://dx.doi.org/10.1016/s0092-8674(03)00040-0]

Mackey, D, Holt, BF, Wiig, A & Dangl, JL (2002) RIN4 interacts with Pseudomonas syringae type III effector molecules and is required for RPM1-mediated resistance in *Arabidopsis. Cell,* 108, 743-54. [http://dx.doi.org/10.1016/s0092-8674(02)00661-x]

Madamanchi, NR & Kuc, J (1991) Induced systemic resistance in plants.*The fungal spore and disease initiation in plants and animals,* Plenum press, New York 347-62.

Mahajan, S, Sopoy, SK & Tuteja, N (2006) CBL-CIPK paradigm: role in calcium and stress signaling in plants. *Proceedings of the Indian National Science Academy,* 72, 63-78.

Malnoy, M, Venisse, JS & Chevreau, E (2005) Expression of a bacterial effector hairpin N causes increased resistance to fire blight in *Pyrus communis. Tree Genet Genomes,* 1, 41-9.
[http://dx.doi.org/10.1007/s11295-005-0006-0]

Mantri, N, Pang, ECK & Ford, R (2010) Molecular Biology for Stress Management.*Climate change and management of cool season grain legume crops,* Springer-Verlag Berlin Heidelberg 377-408.

Manzanares-Dauleux, MJ, Delourme, R, Baron, F & Thomas, G (2000) Mapping of one major gene and of QTLs involved in resistance to clubroot in Brassica napus. *Theor Appl Genet,* 101, 885-91.
[http://dx.doi.org/10.1007/s001220051557]

Marè, C, Mazzucotelli, E, Crosatti, C, Francia, E, Stanca, AM & Cattivelli, L (2004) *Hv-WRKY 38/* a new transcription factor involved in cold- and drought-response in barley. *Plant Mol Biol,* 55, 399-416.
[http://dx.doi.org/10.1007/s11103-004-0906-7]

Marina-García, N, Franchi, L, Kim, Y-G, Miller, D, McDonald, C, Boons, G-J & Núñez, G (2008) Pannexin-1-mediated intracellular delivery of muramyl dipeptide induces caspase-1 activation *via* cryopyrin/NLRP3 independently of Nod2. *J Immunol,* 180, 4050.
[http://dx.doi.org/10.4049/jimmunol.180.6.4050]

Martin, C & Paz-Ares, J (1997) MYB transcription factors in plants. *Trends Genet,* 13, 67-73.
[http://dx.doi.org/10.1016/s0168-9525(96)10049-4]

Martin, GB, Brommonschenkel, SH, Chunwongse, J, Frary, A, Ganal, MW, Spivey, R, Wu, T, Earle, ED & Tanksley, SD (1993) Map-based clonning of a protein kinase gene confering disease resistance in tomato. *Science,* 262, 1432-6.
[http://dx.doi.org/10.1126/science.7902614]

Martins-Salles, S, Machado, V, Massochin-Pinto, L & Fiuza, LM (2017) Genetically modified soybean expressing insecticidal protein (Cry1Ac): Management risk and perspectives. *Facets,* 2, 496-512.
[http://dx.doi.org/10.1139/facets-2017-0006]

Matta, C (2010) Spontaneous generation and disease causation: Anton de Bary's experiments with *Phytophthora infestans* and late blight of potato. *J Hist Biol,* 43, 459-91.
[http://dx.doi.org/10.1007/s10739-009-9220-1]

Maugarny-Calès, A, Gonçalves, B, Jouannic, S, Melkonian, M, Wong, GK-S & Laufs, P (2016) Apparition of the NAC Transcription Factors Predates the Emergence of Land Plants. *Mol Plant,* 9, 1345-8.
[http://dx.doi.org/10.1016/j.molp.2016.05.016]

Mazid, M, Khan, TA & Mohammad, F (2011) Role of secondary metabolites in defense mechanisms of plants. *Biology and Medicine,* 3, 232-49.

McCallum, CM, Comai, L, Greene, EA & Henikoff, S (2000) Targeted screening for induced mutations. *Nat Biotechnol,* 18, 455-7.
[http://dx.doi.org/10.1038/74542]

McDonald, BA & Linde, C (2002) Pathogen population genetics, evolutionary potential and durable resistance. *Annu Rev Phytopathol,* 40, 349-79.

[http://dx.doi.org/10.1146/annurev.phyto.40.120501.101443]

McDowell, JM & Dangl, JL (2000) Signal transduction in the plant immune response. *Trends Biochem Sci,* 25, 79-82.
[http://dx.doi.org/10.1016/S0968-0004(99)01532-7]

McDowell, JM, Dhandaydham, M, Long, TA, Aarts, MG, Goff, S, Holub, EB & Dangl, JL (1998) Intragenic recombination and diversifying selection contribute to the evolution of downy mildew resistance at the *RPP8* locus of *Arabidopsis. Plant Cell,* 10, 1861-74.
[http://dx.doi.org/10.1105/tpc.10.11.1861]

McDowell, JM & Simon, SA (2006) Recent insights into R gene evolution. *Mol Plant Pathol,* 7, 437-48.
[http://dx.doi.org/10.1111/j.1364-3703.2006.00342.x]

McGrann, GRD, Stavrinides, A, Russell, J, Corbitt, MM, Booth, A, Chartrain, L, Thomas, WTB & Brown, JKM (2014) A trade off between mlo resistance to powdery mildew and increased susceptibility of barley to a newly important disease, Ramularia leaf spot. *J Exp Bot,* 65, 1025-37.
[http://dx.doi.org/10.1093/jxb/ert452]

McHale, L, Tan, X, Koehl, P & Michelmore, WR (2006) Plant NBS-LRR proteins: adaptable guards. *Genome Biol,* 7, 212.
[http://dx.doi.org/10.1186/gb-2006-7-4-212]

Medzhitov, R (2001) Toll-like receptors and innate immunity. *Nat Rev Immunol,* 1, 135-45.
[http://dx.doi.org/10.1038/35100529]

Medzhitov, R & Janeway, CA (1997) Innate immunity: The virtues of a nonclonal system of recognition. *Cell,* 91, 295-8.
[http://dx.doi.org/10.1016/s0092-8674(00)80412-2]

Meeley, RB & Walton, JD (1993) Molecular biology and biochemistry of Hm1, a maize gene for fungal resistance. In: Nester, E.W., Verma, D.P.S, (Eds.), *Advances in Molecular Genetics of Plant–Microbe Interactions,* Kluwver Academic Publishers, Dordrecht, The Netherlands 463-7.
[http://dx.doi.org/10.1007/978-94-017-0651-3_51]

Mengiste, T, Chen, X, Salmeron, J & Dietrich, R (2003) The BOTRYTIS SUSCEPTIBLE1 gene encodes an R2R3MYB transcription factor protein that is required for biotic and abiotic stress responses in *Arabidopsis. Plant Cell,* 15, 2551-65.
[http://dx.doi.org/10.1105/tpc.014167]

Meyers, BC, Dickerman, AW, Michelmore, RW, Sivaramakrishnan, S, Sobral, BW & Young, ND (1999) Plant disease resistance genes encode members of an ancient and diverse protein family within the nucleotide-binding superfamily. *Plant J,* 20, 317-32.
[http://dx.doi.org/10.1046/j.1365-313x.1999.t01-1-00606.x]

Meyers, BC, Kozika, A, Griegoa, A, Kuanga, H & Michelmore, RW (2003) Genome-Wide Analysis of NBS-LRR–Encoding Genes in *Arabidopsis. Plant Cell,* 15, 809-34.
[http://dx.doi.org/10.1105/tpc.009308]

Michelmore, RW, Christopoulou, M & Caldwell, KS (2013) Impacts of resistance gene genetics, function, and evolution on a durable future. *Annu Rev Phytopathol,* 51, 291-319.
[http://dx.doi.org/10.1146/annurev-phyto-082712-102334]

Michelmore, RW & Meyers, BC (1998) Clusters of Resistance Genes in Plants Evolve by Divergent Selection and a Birth-and-Death Process. *Genome Res,* 8, 1113-30.

[http://dx.doi.org/10.1101/gr.8.11.1113]

Miklas, PN, Kelly, JD, Beebe, SE & Blair, W (2006) Common bean breeding for resistance against biotic and abiotic stresses: From classical to MAS breeding. *Euphytica,* 147, 105-31.
[http://dx.doi.org/10.1007/s10681-006-4600-5]

Mindrinos, M, Katagiri, F, Yu, GL & Ausubel, FM (1994) The *A. thaliana* disease resistance gene RPS2 encodes a protein containing a nucleotide-binding site and leucine-rich repeats. *Cell,* 78, 1089-99.
[http://dx.doi.org/10.1016/0092-8674(94)90282-8]

Miska, EA & Ferguson-Smith, AC (2016) Transgenerational inheritance: Models and mechanisms of non-DNA sequence-based inheritance. *Science,* 354, 59-63.
[http://dx.doi.org/10.1126/science.aaf4945]

Mittler, R, Vanderauwera, S, Gollery, M & Van Breusegem, F (2004) Reactive oxygen gene network of plants. *Trends Plant Sci,* 9, 490-8.
[http://dx.doi.org/10.1016/j.tplants.2004.08.009]

Miya, A, Albert, P, Shinya, T, Desaki, Y, Ichimura, K, Shirasu, K, Narusaka, Y, Kawakami, N, Kaku, H & Shibuya, N (2007) CERK1, a LysM receptor kinase, is essential for chitin elicitor signaling in *Arabidopsis. Proc Natl Acad Sci USA,* 104, 19613-8.
[http://dx.doi.org/10.1073/pnas.0705147104]

Monaghan, J & Zipfel, C (2012) Plant pattern recognition receptor complexes at the plasma membrane. *Curr Opin Plant Biol,* 15, 349-57.
[http://dx.doi.org/10.1016/j.pbi.2012.05.006]

Mondragón-Palomino, M, Meyers, BC, Michelmore, RW & Gaut, BS (2002) Patterns of positive selection in the complete NBS-LRR gene family of *Arabidopsis thaliana. Genome Res,* 12, 1305-15.
[http://dx.doi.org/10.1101/gr.159402]

Montarry, J, Cartier, E, Jacquemond, M, Palloix, A & Moury, B (2012) Virus adaptation to quantitative plant resistance: erosion or breakdown? *J Evol Biol,* 25, 2242-52.
[http://dx.doi.org/10.1111/j.1420-9101.2012.02600.x]

Montesano, M, Brader, G & Palva, ET (2003) Pathogen derived elicitors: searching for receptors in plants. *Mol Plant Pathol,* 4, 73-9.
[http://dx.doi.org/10.1046/j.1364-3703.2003.00150.x]

Moosa, A, Farzand, A, Sahi, ST & Khan, SA (2018) Transgenic expression of antifungal pathogenesis-related proteins against phytopathogenic fungi – 15 years of success. *Isr J Plant Sci,* 65, 38-54.
[http://dx.doi.org/10.1080/07929978.2017.1288407]

Moradpour, M & Abdulah, SNA (2019) CRISPR/dCas9 Platforms in plants: strategies and applications beyond genome editing. *Plant Biotechnol J,* 18, 32-44.
[http://dx.doi.org/10.1111/pbi.13232]

Moran, NA (1988) The evolution of host-plant alternation in aphids: Evidence for specialization as a dead end. *Am Nat,* 132, 681-706.
[http://dx.doi.org/10.1086/284882]

Morant, AV, Jorgensen, K, Jorgensen, C, Paquette, SM, Sánchez-Pérez, R, Møller, BL & Bak, S (2008) beta-Glucosidases as detonators of plant chemical defense. *Phytochemistry,* 69, 1795-813.
[http://dx.doi.org/10.1016/j.phytochem.2008.03.006]

Mou, Z, Fan, WH & Dong, XN (2003) Inducers of plant systemic acquired resistance regulate NPR1 function through redox changes. *Cell,* 113, 935-44.
[http://dx.doi.org/10.1016/s0092-8674(03)00429-x]

Mundt, CC (2014) Durable resistance: a key to sustainable management of pathogens and pests. *Infect Genet Evol,* 27, 446-55.
[http://dx.doi.org/10.1016/j.meegid.2014.01.011]

Mundt, CC, Cowger, C & Garrett, KA (2002) Relevance of integrated disease management to resistance durability. *Euphytica,* 124, 245-52.
[http://dx.doi.org/10.1023/A:1015642819151]

Mysore, KS & Ryu, C-M (2004) Nonhost resistance: How much do we know? *Trends Plant Sci,* 9, 97-104.
[http://dx.doi.org/10.1016/j.tplants.2003.12.005]

Naito, K, Taguchi, F, Suzuki, T, Inagaki, Y, Toyoda, K, Shiraishi, T & Ichinose, Y (2008) Amino Acid Sequence of Bacterial Microbe-Associated Molecular Pattern flg22 Is Required for Virulence. *Mol Plant Microbe Interact,* 21, 1165-74.
[http://dx.doi.org/10.1094/MPMI-21-9-1165]

Nandris, D, Kohler, F, Monimeau, L & Pellegrin, F (1997) Lutte intégrée contre les ravageurs (Ipm) et approche intégrée du pathosystème Coffea arabica. *ASIC 17ème colloque,* 588-98.https://nph.onlinelibrary.wiley.com/doi/pdf/10.1111/j.1469-8137.2009.02849.x In French

Navaud, O, Barbacci, A, Taylor, A, Clarkson, JP & Raffaele, S (2018) Shifts in diversification rates and host jump frequencies shaped the diversity of host range among Sclerotiniaceae fungal plant pathogens. *Mol Ecol,* 27, 1309-23.
[http://dx.doi.org/10.1111/mec.14523]

Nekrasov, V, Wang, C, Win, J, Lanz, C, Weigel, D & Kamoun, S (2017) Rapid generation of a transgene-free powdery mildew resistant tomato by genome deletion. *Sci Rep,* 7, 482.
[http://dx.doi.org/10.1038/s41598-017-00578-x]

Nekrutenko, A & Li, WH (2001) Transposable elements are found in a large number of human protein-coding genes. *Trends Genet,* 17, 619-21.
[http://dx.doi.org/10.1016/s0168-9525(01)02445-3]

Ngugi, HK, King, SB, Holt, J & Julian, M (2001) Simultaneous temporal progress of sorghum anthracnose and leaf blight in crop mixtures with disparate patterns. *Phytopathology,* 91, 720-9.
[http://dx.doi.org/10.1094/PHYTO.2001.91.8.720]

Niculaes, C, Abramov, A, Hannemann, L & Frey, M (2018) Plant Protection by Benzoxazinoids—Recent Insights into Biosynthesis and Function. *Agronomy (Basel),* 8, 143.
[http://dx.doi.org/10.3390/agronomy8080143]

Niehl, A, Wyrsch, I, Boller, T & Heinlein, M (2016) Double-stranded RNAs induce a pattern-triggered immune signaling pathway in plants. *New Phytol,* 211, 1008-19.
[http://dx.doi.org/10.1111/nph.13944]

Niks, RE & Marcel, TC (2009) Nonhost and basal resistance: how to explain specificity? *New Phytol,* 182, 817-28.
[http://dx.doi.org/10.1111/j.1469-8137.2009.02849.x]

Niks, RE, Qi, XQ & Marcel, TC (2015) Quantitative resistance to biotrophic filamentous plant pathogens:

concepts, misconceptions and mechanisms.*Annual Review of Phytopathology,* 445-70.

Niks, RE, Qi, XQ & Marcel, TC (2015) Quantitative resistance to biotrophic filamentous plant pathogens: concepts, misconceptions and mechanisms.*Annual Review of Phytopathology,* Annual Reviews, Palo Alto, CA 445-70.

Niks, RE, Qi, XQ & Marcel, TC (2015) Quantitative resistance to biotrophic filamentous plant pathogens: concepts, misconceptions and mechanisms. *Annual Review of Phytopathology,* 53, 445-70.https://doi.org/110.1146/annurev-phyto-080614-115928

Nishimura, MT, Monteiro, F & Dangl, JL (2015) Treasure your exceptions: unusual domains in immune receptors reveal host virulence targets. *Cell,* 161, 957-60.
[http://dx.doi.org/10.1016/j.cell.2015.05.017]

Niu, QW, Lin, SS, Reyes, JL, Chen, KC, Wu, HW, Yeh, SD & Chua, NH (2006) Expression of artificial microRNAs in transgenic Arabidopsis thaliana confers virus resistance. *Nature Biotechnol,* 24, 1420-8.https://doi.org/110.1038/nbt1255

Noir, S (2002) Diversité des gènes de résistance au sein du génome des caféiers (Coffea L.). *Analyse génétique de la résistance au nématode à galles, Meloidogyne exigua chez C Arabica,* PhD Thesis. University of Montpellier II, France 166.

Nürnberger, T & Brunner, F (2002) Innate immunity in plants and animals: emerging parallels between the recognition of general elicitors and pathogen-associated molecular patterns. *Curr Opin Plant Biol,* 5, 318-24.
[http://dx.doi.org/10.1016/s1369-5266(02)00265-0]

Nürnberger, T, Brunner, F, Kemmerling, B & Piater, L (2004) Innate immunity in plants and animals: Striking similarities and obvious differences. *Immunol Rev,* 198, 249-66.
[http://dx.doi.org/10.1111/j.0105-2896.2004.0119.x]

Nürnberger, T & Lipka, V (2005) Non-host resistance in plants: new insights into an old phenomenon. *Mol Plant Pathol,* 6, 335-45.
[http://dx.doi.org/10.1111/j.1364-3703.2005.00279.x]

Nuruzzaman, M, Sharoni, AM, Satoh, K, Karim, MR, Harikrishna, JA, Shimizu, T, Sasaya, T, Omura, T, Haque, MA, Hasan, SM, Ahmad, A & Kikuchi, S (2015) NAC transcription factor family genes are differentially expressed in rice during infections with Rice dwarf virus, Rice black-streaked dwarf virus, Rice grassy stunt virus, Rice ragged stunt virus, and Rice transitory yellowing virus. *Front Plant Sci,* 6, 676.
[http://dx.doi.org/10.3389/fpls.2015.00676]

Obrycki, JJ, Losey, JE, Taylor, OR & Jesse, LCH (2001) Transgenic Insecticidal Corn: Beyond Insecticidal Toxicity to Ecological Complexity. *Bioscience,* 51, 353-61.
[http://dx.doi.org/10.1641/0006-3568(2001)051[0353:TICBIT]2.0.CO;2]

Olsen, AN, Ernst, HA, Leggio, LL & Skriver, K (2005) NAC Transcription Factors: Structurally Distinct, Functionally Diverse. *Trends Plant Sci,* 10, 79-87.
[http://dx.doi.org/10.1016/j.tplants.2004.12.010]

Onaga, G & Wydra, K (2016) Advances in plant tolerance to biotic stresses.*Plant Genomics,* IntechOpen, Rijeka, Croatia 229-72.

Onaga, G & Wydra, K (2016) Advances in Plant Tolerance to Biotic Stresses.*Plant Genomics,* IntechOpen, Rijeka, Croatia 229-72.

(2019) One thousand plant transcriptomes and the phylogenomics of green plants. *Nature,* 574, 679-85.

[http://dx.doi.org/10.1038/s41586-019-1693-2]

Ongena, M, Daayf, F, Jacques, P, Thonart, P, Benhamou, N, Paulitz, TC & Belanger, RR (2000) Systemic induction of phytoalexins in cucumber in response to treatments with fluorescent pseudomonads. *Plant Pathol,* 49, 523-30.
[http://dx.doi.org/10.1046/j.1365-3059.2000.00468.x]

Opdenakker, K, Remans, T, Vangronsveld, J & Cuypers, A (2012) Mitogen-Activated Protein (MAP) kinases in plant metal stress, regulation and responses in comparison to other biotic and abiotic stresses. *Int J Mol Sci,* 13, 7828-53.
[http://dx.doi.org/10.3390/ijms13067828]

Osuna-Cruz, CM, Paytuvi-Gallart, A, Di Donato, A, Sundesha, V, Andolfo, G, Aiese Cigliano, R, Sanseverino, W & Ercolano, MR (2018) PRGdb 3.0: A comprehensive platform for prediction and analysis of plant disease resistance genes. *Nucleic Acids Res,* 46, D1197-201.
[http://dx.doi.org/10.1093/nar/gkx1119]

Paccanaro, MC, Sella, L, Castiglioni, C, Giacomello, F, Martínez-Rocha, AL, D'Ovidio, R, Schäfer, W & Favaron, F (2017) Synergistic Effect of Different Plant Cell Wall-Degrading Enzymes Is Important for Virulence of *Fusarium graminearum. Mol Plant Microbe Interact,* 30, 886-95.
[http://dx.doi.org/10.1094/MPMI-07-17-0179-R]

Padmanabhan, (1973) The great Bengal famine. *Annual Review of Phytopathology,* 11, 11-24.

Pal, KK & Gardener, BMS (2006) Biological Control of Plant Pathogens. *The Plant Health Instructor.*
[http://dx.doi.org/10.1094/PHI-A-2006-1117-02]

Palloix, A, Ayme, V & Moury, B (2009) Durability of plant major resistance genes to pathogens depends on the genetic background, experimental evidence and consequences for breeding strategies. *New Phytol,* 183, 190-9.
[http://dx.doi.org/10.1111/j.1469-8137.2009.02827.x]

Palloix, A & Ordon, F (2011) Advanced breeding for virus resistance in plants.*Recent Advances in Plant Virology,* Caister Academic Press, Norwich 195-218.

Pan, L, Zhang, X, Wang, J, Ma, X, Zhou, M, Huang, LK, Nie, G, Wang, P, Yang, Z & Li, J (2016) Transcriptional profiles of drought-related genes in modulating metabolic processes and antioxidant defenses in *Lolium multiflorum. Front Plant Sci,* 7, 519.
[http://dx.doi.org/10.3389/fpls.2016.00519]

Pan, Q, Wendel, J & Fluhr, R (2000) Divergent evolution of plant NBS-LRR resistance gene homologues in dicot and cereal genomes. *J Mol Evol,* 50, 203-13.
[http://dx.doi.org/10.1007/s002399910023]

Panda, N & Khush, GS (1995) *Host Plant Resistance to Insects Wallingford*CABI/IRRI, UK.

Pande, A, Saxena, SC, Thapliyal, M, Guru, SK, Kumar, A & Arora, S (2018) Role of AP2/EREBP Transcription Factor Family in Environmental Stress Tolerance. *Cell & Cellular Life Sciences Journal,* 3, 000120.
[http://dx.doi.org/10.23880/CCLSJ-16000120]

Park, CY, Lee, JH, Yoo, JH, Moon, BC, Choi, MS, Kang, YH, Lee, SM, Kim, HS, Kang, KY, Chung, WS, Lim, CO & Cho, MJ (2005) WRKY group IId transcription factors interact with calmodulin. *FEBS Lett,* 579, 1545-50.

[http://dx.doi.org/10.1016/j.febslet.2005.01.057]

Parker, J (2001) *Elongation Factors,* Translation. Encyclopedia of Genetics 610-1.
[http://dx.doi.org/10.1006/rwgn.2001.0402]

Parker, JE (2003) Plant recognition of microbial patterns. *Trends Plant Sci,* 8, 245-7.
[http://dx.doi.org/10.1016/S1360-1385(03)00105-5]

Parlevliet, JE (2002) Durability of resistance against fungal, bacterial and viral pathogens; present situation. *Euphytica,* 124, 147-56.
[http://dx.doi.org/10.1023/a:1015601731446]

Parolin, P, Bresch, C, Desneux, N, Brun, R, Bout, A, Boll, R & Poncet, C (2012) Secondary plants used in biological control: A review. *Int J Pest Manage,* 58, 91-100.
[http://dx.doi.org/10.1080/09670874.2012.659229]

Paulitz, TC, Zhou, T & Rankin, L (1992) Selection of rhizosphere bacteria for biological control of *Pythium aphanidermatum* on hydroponically grown cucumber. *Biol Control,* 2, 226-37.
[http://dx.doi.org/10.1016/1049-9644(92)90063-j]

Pavet, V, Quintero, C, Cecchini, NM, Rosa, AL & Alvarez, ME (2006) *Arabidopsis* displays centromeric DNA hypomethylation and cytological alterations of heterochromatin upon attack by *Pseudomonas syringae.* *Mol Plant Microbe Interact,* 19, 577-87.
[http://dx.doi.org/10.1094/MPMI-19-0577]

Paz-Ares, J, Ghosal, D, Wienand, U, Peterson, PA & Saedler, H (1987) The regulatory c1 locus of Zea mays encodes a protein with homology to myb proto-oncogene products and with structural similarities to transcriptional activators. *EMBO J,* 6, 3553-8.
[http://dx.doi.org/10.1002/j.1460-2075.1987.tb02684.x]

Pechan, T, Jiang, B, Steckler, D, Ye, L, Lin, L, Luthe, DS & Williams, WP (1999) Characterization of three distinct cDNA clones encoding cysteine proteinases from maize (*Zea mays* L.) callus. *Plant Mol Biol,* 40, 111-9.
[http://dx.doi.org/10.1023/A:1026494813936]

Pereira, JF & Ryan, PR (2019) The Role of Transposable Elements in the Evolution of Aluminium Resistance in Plants. *J Exp Bot,* 70, 41-54.
[http://dx.doi.org/10.1093/jxb/ery357]

Peressotti, E, Wiedemann-Merdinoglu, S, Delmotte, F, Bellin, D, Di Gaspero, G, Testolin, R, Merdinoglu, D & Mestre, P (2010) Breakdown of resistance to grapevine downy mildew upon limited deployment of a resistant variety. *BMC Plant Biol,* 10, 147.
[http://dx.doi.org/10.1186/1471-2229-10-147]

Petersen, M, Brodersen, P, Naested, H, Andreasson, E, Lindhart, U, Johansen, B, Nielsen, HB, Lacy, M, Austin, MJ, Parker, JE, Sharma, SB, Klessig, DF, Martienssen, R, Mattsson, O, Jensen, AB & Mundy, J (2000) *Arabidopsis* map kinase 4 negatively regulates systemic acquired resistance. *Cell,* 103, 1111-20.
[http://dx.doi.org/10.1016/s0092-8674(00)00213-0]

Piedras, P, Rivas, S, Dröge, S, Hillmer, S & Jones, JDG (2000) Functionnal, c- myctagged *Cf-9* resistance gene products are plasma-membrane localized and glycophosphorylated. *Plant J,* 21, 529-36.
[http://dx.doi.org/10.1046/j.1365-313x.2000.00697.x]

Pieterse, CM, Leon-Reyes, A, Van der Ent, S & Van Wees, SCM (2009) Networking by small-molecule hormones in plant immunity. *Nat Chem Biol,* 5, 308-16.

[http://dx.doi.org/10.1038/nchembio.164]

Pieterse, CM, Van der Does, D, Zamioudis, C, Leon-Reyes, A & Van Wees, SC (2012) Hormonal modulation of plant immunity. *Annu Rev Cell Dev Biol,* 28, 489-521.
[http://dx.doi.org/10.1146/annurev-cellbio-092910-154055]

Pieterse, CMJ & Van Loon, L (2004) NPR1: the spider in the web of induced resistance signaling pathways. *Curr Opin Plant Biol,* 7, 456-64.
[http://dx.doi.org/10.1016/j.pbi.2004.05.006]

Pieterse, CMJ, Zamioudis, C, Berendsen, RL, Weller, DM, Van Wees, SCM & Bakker, PAHM (2014) Induced Systemic Resistance by Beneficial Microbes. *Annu Rev Phytopathol,* 52, 347-75.
[http://dx.doi.org/10.1146/annurev-phyto-082712-102340]

Pietravalle, S, Lemarie, S & van den Bosch, F (2006) Durability of resistance and cost of virulence. *Eur J Plant Pathol,* 114, 107-16.
[http://dx.doi.org/10.1007/s10658-005-3479-7]

Pilet-Nayel, ML, Moury, B, Caffier, V, Montarry, J, Kerlan, MC, Fournet, S, Durel, CE & Delourme, R (2017) Quantitative resistance to plant pathogens in pyramiding strategies for durable crop protection. *Front Plant Sci,* 8, 1838.
[http://dx.doi.org/10.3389/fpls.2017.01838]

Piriyapongsa, J, Mariño-Ramírez, L & Jordan, IK (2007) Origin and evolution of human microRNAs from transposable elements. *Genetics,* 176, 1323-37.
[http://dx.doi.org/10.1534/genetics.107.072553]

Plasterk, RHA (2002) RNA silencing: the genome's immune system. *Science,* 296, 1263-5.
[http://dx.doi.org/10.1126/science.1072148]

Platt, A, Gugger, PF, Pellegrini, M & Sork, VL (2015) Genome-wide signature of local adaptation linked to variable CpG methylation in oak populations. *Mol Ecol,* 24, 3823-30.
[http://dx.doi.org/10.1111/mec.13230]

Poland, J & Rutkoski, J (2016) Advances and challenges in genomic selection for disease resistance. In: Leach, J.E., Lindow, S., (Eds.), *Annual Review of Phytopathology,* Palo Alto, CA 79-98.

Poland, JA, Balint-Kurti, PJ, Wisser, RJ, Pratt, RC & Nelson, RJ (2009) Shades of gray: the world of quantitative disease resistance. *Trends Plant Sci,* 14, 21-9.
[http://dx.doi.org/10.1016/j.tplants.2008.10.006]

Porter, BW, Paidi, M, Ming, R, Alam, M, Nishijima, WT & Zhu, YJ (2009) Genome-wide analysis of *Carica papaya* reveals a small NBS resistance gene family. *Mol Genet Genomics,* 281, 609-26.
[http://dx.doi.org/10.1007/s00438-009-0434-x]

Potnis, N, Branham, SE, Jones, JB & Wechter, WP (2019) Genome-wide association study of resistance to *Xanthomonas gardneri* in the USDA pepper (*Capsicum*) collection. *Phytopathology,* 109, 1217-25.
[http://dx.doi.org/10.1094/PHYTO-06-18-0211-R]

Poulin, R & Keeney, DB (2008) Host specificity under molecular and experimental scrutiny. *Trends Parasitol,* 24, 24-8.
[http://dx.doi.org/10.1016/j.pt.2007.10.002]

Prasch, CM & Sonnewald, U (2013) Simultaneous application of heat, drought and virus to Arabidopsis plants reveals significant shifts in signaling networks. *Plant Physiol,* 162, 1849-66.

[http://dx.doi.org/10.1104/pp.113.221044]

Prince, DC, Drurey, C, Zipfel, C & Hogenhout, SA (2014) The leucine-rich repeat receptor-like kinase brassinosteroid insensitive1-associated kinase1 and the cytochrome p450 phytoalexin deficient3 contribute to innate immunity to aphids in Arabidopsis. *Plant Physiol,* 164, 2207-19.
[http://dx.doi.org/10.1104/pp.114.235598]

Prioul, S, Frankewitz, A & Deniot, G (2004) Mapping of quantitative trait loci for partial resistance to *Mycosphaerella pinodes* in pea (*Pisum sativum* L.), at the seedling and adult plant stages. *Theor Appl Genet,* 108, 1322-34.
[http://dx.doi.org/10.1007/s00122-003-1543-2]

Pritham, EJ, Putliwala, T & Feschotte, C (2007) Mavericks, a novel class of giant transposable elements widespread in eukaryotes and related to DNA viruses. *Gene,* 390, 3-17.
[http://dx.doi.org/10.1016/j.gene.2006.08.008]

Pryor, T & Ellis, JG (1993) The genetic complexity of fungal resistance genes in plants. *Adv Plant Pathol,* 10, 282-305.

Pulliam, DA, Williams, DL & Stewart, CN (2001) Isolation and characterization of a serine protenase inhibitor cDNA from cabbage and its antibiosis in transgenic tobacco plants. *Plant Cell Biotechnol Mol Biol,* 2, 19-32.

Pyott, DE & Molnar, A (2015) Going mobile: non-cell-autonomous small RNAs shape the genetic landscape of plants. *Plant Biotechnol J,* 13, 306-18.
[http://dx.doi.org/10.1111/pbi.12353]

Qiao, F (2015) Fifteen Years of *Bt* Cotton in China: The Economic Impact and its Dynamics. *World Dev,* 70, 177-85.
[http://dx.doi.org/10.1016/j.worlddev.2015.01.011]

Qiu, Y, Guo, J, Jing, S, Zhu, L & He, G (2010) High resolution mapping of the brown planthopper resistance gene *Bph6* in rice and characterizing its resistance in the 9311 and Nipponbare near isogenic backgrounds. *Theor Appl Genet,* 121, 1601-11.
[http://dx.doi.org/10.1007/s00122-010-1413-7]

Qu, F, Ren, T & Morris, TJ (2003) The coat protein of Turnip crinkle virus suppresses post transcriptional gene silencing at an early initiation step. *J Virol,* 77, 511-22.
[http://dx.doi.org/10.1128/JVI.77.1.511-522.2003]

Quadrana, L, Bortolini Silveira, A, Mayhew, GF, LeBlanc, C, Martienssen, RA, Jeddeloh, JA & Colot, V (2016) The Arabidopsis thaliana mobilome and its impact at the species level (D Zilberman, Ed.).*eLife,* 5, e15716.

Quenouille, J, Montarry, J, Palloix, A & Moury, B (2013) Farther, slower, stronger: how the plant genetic background protects a major resistance gene from breakdown. *Mol Plant Pathol,* 14, 109-18. a
[http://dx.doi.org/10.1111/j.1364-3703.2012.00834.x]

Quenouille, J, Paulhiac, E, Moury, B & Palloix, A (2014) Quantitative trait loci from the host genetic background modulate the durability of a resistance gene: a rational basis for sustainable resistance breeding in plants. *Heredity,* 112, 579-87.
[http://dx.doi.org/10.1038/hdy.2013.138]

Quenouille, J, Vassilakos, N & Moury, B (2013) A major crop pathogen that has provided major insights into the evolution of viral pathogenicity. *Mol Plant Pathol,* 14, 439-52. b

[http://dx.doi.org/10.1111/mpp.12024]

Quispe-Huamanquispe, DG, Gheysen, G & Kreuze, JF (2017) Horizontal Gene Transfer Contributes to Plant Evolution: The Case of *Agrobacterium* T-DNAs. *Front Plant Sci,* 8, 2015.
[http://dx.doi.org/10.3389/fpls.2017.02015]

Rahbe, Y, Deraison, C, Bonadé-Bottino, M, Girard, C, Nardon, C & Jouanin, L (2003) Effects of the cysteine protease inhibitor oryzacystatin (OC-I) on different aphids and reduced performance of *Myzus persicae* on OC-I expressing transgenic oilseed rape. *Plant Sci,* 164, 441-50.
[http://dx.doi.org/10.1016/S0168-9452(02)00402-8]

Rahmatov, M, Otambekova, M & Muminjanov, H (2019) Characterization of stem, stripe and leaf rust resistance in Tajik bread wheat accessions. *Euphytica,* 215, 55.
[http://dx.doi.org/10.1007/s10681-019-2377-6]

Rahnama, H, Nikmard, M, Abolhasani, M, Osfoori, R, Sanjarian, F & Habashi, AA (2017) Immune analysis of cry1Ab-genetically modified potato by in-silico analysis and animal model. *Food Sci Biotechnol,* 26, 1437-45.
[http://dx.doi.org/10.1007/s10068-017-0181-4]

Ramamoorthy, R, Jiang, SY, Kumar, N, Venkatesh, PN & Ramachandran, S (2008) A comprehensive transcriptional profiling of the WRKY gene family in rice under various abiotic and phytohormone treatments. *Plant Cell Physiol,* 49, 865-79.
[http://dx.doi.org/10.1093/pcp/pcn061]

Ramamoorthy, V, Viswanathan, R, Raguchander, T, Prakasam, V & Samiyappan, R (2001) Induction of systemic resistance by plant growth promoting rhizobacteria in crop plants against pests and diseases. *Crop Prot,* 20, 1-11.
[http://dx.doi.org/10.1016/S0261-2194(00)00056-9]

Rambani, A, Rice, JH, Liu, J, Lane, T, Ranjan, P, Mazarei, M, Pantalone, V, Stewart, CN, Staton, M & Hewezi, T (2015) The methylome of soybean roots during the compatible interaction with the soybean cyst nematode. *Plant Physiol,* 168, 1364.
[http://dx.doi.org/10.1104/pp.15.00826]

Ramos, HC, Rumbo, M & Sirard, JC (2004) Bacterial flagellins: Mediators of pathogenicity and host immune responses in mucosa. *Trends Microbiol,* 12, 509-17.
[http://dx.doi.org/10.1016/j.tim.2004.09.002]

Reeck, GR (1997) Proteinase inhibitors and resistance of transgenic plants to insects.*Advance in insect control: the role of transgenic plants,* Taylor and Francis Press, London 157-83.

Reiss, ER & Drinkwater, LE (2018) Cultivar mixtures: a meta-analysis of the effect of intraspecific diversity on crop yield. *Ecol Appl,* 28, 62-77.
[http://dx.doi.org/10.1002/eap.1629]

Resende, DC, Mendes, SM, Marucci, RC, Silva, AC, Campanha, MM & Waquil, JM (2016) Does Bt maize cultivation affect the non-target insect community in the agro ecosystem? *Rev Bras Entomol,* 60, 82-93.
[http://dx.doi.org/10.1016/j.rbe.2015.12.001]

Restrepo-Montoya, D, Brueggeman, R, McClean, PE & Osorno, JM (2020) Computational identification of receptor-like kinases RLK and receptor-like proteins RLP in legumes. *BMC Genomics,* 21, 459.
[http://dx.doi.org/10.1186/s12864-020-06844-z]

Rey, O, Danchin, E, Mirouze, M, Loot, C & Blanchet, S (2016) Adaptation to Global Change: A

Transposable Element–Epigenetics Perspective. *Trends Ecol Evol,* 31, 514-26. [http://dx.doi.org/10.1016/j.tree.2016.03.013]

Ribeiro, APO, Pereira, EJG, Galvan, TL, Picanço, MC, Picoli, EAT, Da Silva, DJH, Fári, MG & Otoni, WC (2006) Effect of eggplant transformed with oryzacystatin gene on *Myzus persicae* and *Macrosiphon euphorbiae. J Appl Entomol,* 130, 84-90. [http://dx.doi.org/10.1111/j.1439-0418.2005.01021.x]

Ricci, P, Bui, S & Lamine, C (2011) *Repenser la protection des cultures: innovations et transitions,* Editions Quae 250. [In French]

Richly, E, Kurth, J & Leister, D (2002) Mode of amplification and reorganization of resistance genes during recent *Arabidopsis thaliana* evolution. *Mol Biol Evol,* 19, 76-84. [http://dx.doi.org/10.1093/oxfordjournals.molbev.a003984]

Richter, TE & Ronald, PC (2000) The evolution of disease resistance genes. *Plant Mol Biol,* 42, 195-204.

Riechmann, JL, Heard, J, Martin, G, Reuber, L, Jiang, C, Keddie, J, Adam, L, Pineda, O, Ratcliffe, OJ, Samaha, RR, Creelman, R, Pilgrim, M, Broun, P, Zhang, JZ, Ghandehari, D, Sherman, BK & Yu, G (2000) *Arabidopsis* transcription factors: genome-wide comparative analysis among eukaryotes. *Science,* 290, 2105-10. [http://dx.doi.org/10.1126/science.290.5499.2105]

Robatzek, S, Bittel, P, Chinchilla, D, Kochner, P, Felix, G, Shiu, SH & Boller, T (2007) Molecular identification and characterization of the tomato flagellin receptor LeFLS2, an orthologue of Arabidopsis FLS2 exhibiting characteristically different perception specificities. *Plant Mol Biol,* 64, 539-47. [http://dx.doi.org/10.1007/s11103-007-9173-8]

Robert, V & Bucheton, A (2004) Régulation de l'expression des séquences répétées et interférence par l'ARN. *M/S: médecine sciences,* 20, 767-72.https://id.erudit.org/iderudit/008980ar

Robert-Seilaniantz, A, MacLean, D, Jikumaru, Y, Hill, L, Yamaguchi, S, Kamiya, S & Jones, JDG (2011) The microRNA miR393 Re-Directs Secondary Metabolite Biosynthesis Away From Camalexin and Towards Glucosinolates. *Plant J,* 67, 218-31. [http://dx.doi.org/10.1111/j.1365-313X.2011.04591.x]

Robert-Seilaniantz, A, Navarro, L, Bari, R & Jones, JD (2007) Pathological hormone imbalances. *Curr Opin Plant Biol,* 10, 372-9. [http://dx.doi.org/10.1016/j.pbi.2007.06.003]

Rocherieux, J, Glory, P, Giboulot, A, Boury, S, Barbeyron, G, Thomas, G & Manzanares-Dauleux, MJ (2004) Isolate-specific and broad-spectrum QTLs are involved in the control of clubroot in Brassica oleracea. *Theor Appl Genet,* 108, 1555-63. [http://dx.doi.org/10.1007/s00122-003-1580-x]

Rodríguez López, CM & Wilkinson, MJ (2015) Epi-fingerprinting and epi-interventions for improved crop production and food quality. *Front Plant Sci,* 6, 397. [http://dx.doi.org/10.3389/fpls.2015.00397]

Römer, P, Recht, S & Lahaye, T (2009) A single plant resistance gene promoter engineered to recognize multiple TAL effectors from disparate pathogens. *Proc Natl Acad Sci USA,* 106, 20526-31. [http://dx.doi.org/10.1073/pnas.0908812106]

Ron, M & Avni, A (2004) The receptor for the fungal elicitor ethylene-inducing xylanase is a member of a resistance-like gene family in tomato. *Plant Cell,* 16, 1604-15.

[http://dx.doi.org/10.1105/tpc.022475]

Rooney, HCE, van't Klooster, JW, van der Hoorn, RAL, Joosten, MHAJ & Jones, JDG (1783–1786) Wit PJGMde (2005) Cladosporium Avr2 Inhibits Tomato Rcr3 Protease Required for Cf-2-Dependent Disease Resistance. *Science,* 308
[http://dx.doi.org/10.1126/science.1111404]

Ross, AF (1961) Localized acquired resistance to plant virus infection in hypersensitive hosts. *Virology,* 14, 329-39. a
[http://dx.doi.org/10.1016/0042-6822(61)90318-x]

Ross, AF (1961) Systemic acquired resistance induced by localized virus infections in plants. *Virology,* 14, 340-58. b
[http://dx.doi.org/10.1016/0042-6822(61)90319-1]

Rouxel, T, Penaud, A & Pinochet, X (2003) A 10-year Survey of Populations of *Leptosphaeria maculans* in France Indicates a Rapid Adaptation Towards the *Rlm1* Resistance Gene of Oilseed Rape. *Eur J Plant Pathol,* 109, 871-81.
[http://dx.doi.org/10.1023/A:1026189225466]

Ruggieri, V, Nunziata, A & Barone, A (2014) Positive selection in the leucine-rich repeat domain of *Gro1* genes in *Solanum* species. *J Genet,* 93, 755-65.
[http://dx.doi.org/10.1007/s12041-014-0458-9]

Rushton, PJ, Somssich, IE, Ringler, P & Shen, QJ (2010) WRKY transcription factors. *Trends Plant Sci,* 15, 247-58.
[http://dx.doi.org/10.1016/j.tplants.2010.02.006]

Russel, LM (1977) Hosts and distribution of the greenhouse whitefly, Trialeurodes vaporariorum (Westwood) (Hemiptera: Homoptera: A1- eyrodidae). *United States Department of Agriculture Cooperative Plant Pest Report,* 2, 449-58.

Rustgi, S, Boex-Fontvieille, E, Reinbothe, C, von Wettstein, D & Reinbothe, S (2017) The complex world of plant protease inhibitors: Insights into a Kunitz-type cysteine protease inhibitor of *Arabidopsis thaliana. Commun Integr Biol,* 11, e1368599.
[http://dx.doi.org/10.1080/19420889.2017.1368599]

Ryan, CA (1990) Protease inhibitors in plants: genes for improving defenses against insects and pathogens. *Annu Rev Phytopathol,* 28, 425-49.

Sánchez-Sánchez, H & Morquecho-Contreras, A (2017) Chemical plant defense against herbivores. In: Vonnie, D.C., Shields, , (Eds.), *Herbivores,* IntechOpen, Rijeka, Croatia 3-28.
[http://dx.doi.org/10.5772/67346]

Saker, MM, Salama, HS, Salama, M, El-Banna, A & Abdel Ghany, NM (2011) Production of transgenic tomato plants expressing Cry2Ab gene for the control of some lepidopterous insects endemic in Egypt. *Journal of Genetic Engineering and Biotechnology,* 9, 149-55.
[http://dx.doi.org/10.1016/j.jgeb.2011.08.001]

Sakuma, Y, Liu, Q, Dubouzet, JG, Abe, H, Shinozaki, K & Yamaguchi-Shinozaki, K (2002) DNA-binding specificity of the AP2/ERF domain of Arabidopsis DREBs, transcription factors involved in dehydration-and cold-inducible gene expression. *Biochem Biophys Res Commun,* 290, 998-1009.
[http://dx.doi.org/10.1006/bbrc.2001.6299]

Saijo, Y & Loo, EP (2020) Plant immunity in signal integration between biotic and abiotic stress responses.

New Phytologist, 225, 87-104.
[http://dx.doi.org/10.1111/nph.15989]

Samanani, N & Facchini, PJ (2006) Chapter Three - Compartmentalization of Plant Secondary Metabolism. *Recent Adv Phytochem,* 40, 53-83.
[http://dx.doi.org/10.1016/S0079-9920(06)80037-7]

Sanger, F, Air, GM, Barrell, BG & Brown, NL (1977) Nucleotide Sequence of Bacteriophage Phi X174 DNA. *Nature,* 265, 687-95.
[http://dx.doi.org/10.1038/265687a0]

Sanseverino, W & Ercolano, MR (2012) *In silico* approach to predict candidate R proteins and to define their domain architecture. *BMC Res Notes,* 5, 678.
[http://dx.doi.org/10.1186/1756-0500-5-678]

Sanseverino, W, Roma, G, De Simone, M, Faino, L, Melito, S, Stupka, E, Frusciante, L & Ercolano, MR (2010) PRGdb: a bioinformatics platform for plant resistance gene analysis. *Nucleic Acids Res,* 38, D814-21.
[http://dx.doi.org/10.1093/nar/gkp978]

Sarris, PF, Cevik, V, Dagdas, G, Jones, JDG & Krasileva, KV (2016) Comparative analysis of plant immune receptor architectures uncovers host proteins likely targeted by pathogens. *BMC Biol,* 14, 8.
[http://dx.doi.org/10.1186/s12915-016-0228-7]

Sathoff, AE, Velivelli, S, Shah, DM & Samac, DA (2019) Plant Defensin Peptides have Antifungal and Antibacterial Activity Against Human and Plant Pathogens. *Disease Control and Pest Management,* 109, 402-8.
[http://dx.doi.org/10.1094/PHYTO-09-18-0331-R]

Sathoff, AE, Velivelli, S, Shah, DM & Samac, DA (2019) Plant Defensin Peptides Have Antifungal and Antibacterial Activity Against Human and Plant Pathogens. *Phytopathology,* 109, 402-8.
[http://dx.doi.org/10.1094/PHYTO-09-18-0331-R]

Saucet, SB, Ma, Y, Sarris, PF, Furzer, OJ, Sohn, KH & Jones, JD (2015) Two linked pairs of Arabidopsis TNL resistance genes independently confer recognition of bacterial effector AvrRps4. *Nat Commun,* 6, 6338.
[http://dx.doi.org/10.1038/ncomms7338]

Sauer, M & Friml, J (2010) Immunolocalization of proteins in plants. *Methods Mol Biol,* 655, 253-63.
[http://dx.doi.org/10.1007/978-1-60761-765-5_17]

Savary, S, Willocquet, L, Pethybridge, SJ, Esker, P, McRoberts, N & Nelson, A (2019) The global burden of pathogens and pests on major food crops. *Nat Ecol Evol,* 3, 430-9.
[http://dx.doi.org/10.1038/s41559-018-0793-y]

Savirnata, NM, Jukunen-Titto, R & Oksanen, E (2010) Leaf phenolic compounds in red clover (*Trifolium pratense* L.) induced by exposure to moderately elevated ozone. *Environ Pollut,* 158, 440-6.
[http://dx.doi.org/10.1016/j.envpol.2009.08.029]

Scheffer, RJ (1983) Biological control of Dutch elm disease by Pseudomonas species. *Ann Appl Biol,* 103, 21-30.
[http://dx.doi.org/10.1111/j.1744-7348.1983.tb02736.x]

Schenk, U, Westendorf, AM, Radaelli, E, Casati, A, Ferro, M, Fumagalli, M, Verderio, C, Buer, J, Scanziani, E & Grassi, F (2008) Purinergic control of T cell activation by ATP released through pannexin-1 hemichannels. *Sci Signal,* 1, ra6.

[http://dx.doi.org/10.1126/scisignal.1160583]

Schimoler-O'Rourke, R, Richardson, M & Selitrennikoff, CP (2001) Zeamatin inhibits trypsin and_-amylase activities. *Appl Environ Microbiol,* 67, 2365-6.
[http://dx.doi.org/10.1128/AEM.67.5.2365-2366.2001]

Schlichting, CD & Wund, MA (2014) Phenotypic Plasticity and Epigenetic Marking: An Assessment of Evidence for Genetic Accommodation. *Evolution,* 68, 656-72.
[http://dx.doi.org/10.1111/evo.12348]

Schoeneweiss, DF (1975) Predisposition, stress, and plant disease. *Annu Rev Phytopathol,* 13, 193-211.

Schoonbeek, HJ, Wang, HH, Stefanato, FL, Craze, M, Bowden, S, Wallington, E, Zipfel, C & Ridout, CJ (2015) *Arabidopsis* EF-Tu receptor enhances bacterial disease resistance in transgenic wheat. *New Phytol,* 206, 606-13.
[http://dx.doi.org/10.1111/nph.13356]

Schornack, S, Huitema, E, Cano, LM & Bozkurt, TO (2009) Ten things to know about oomycete effectors. *Mol Plant Pathol,* 10, 795-803.
[http://dx.doi.org/10.1111/j.1364-3703.2009.00593.x]

Schottens-Toma, IMJ & DeWit, PJM (1988) Purification and primary structure of a necrosis-inducing peptide from the apoplastic fluids of tomato infected with *Cladosporium fulvum* (syn. *Fulvia fulva*). *Physiol Mol Plant Pathol,* 33, 59-67.
[http://dx.doi.org/10.1016/0885-5765(88)90043-4]

Schroder, K & Tschopp, J (2010) The inflammasomes. *Cell,* 140, 821-32.
[http://dx.doi.org/10.1016/j.cell.2010.01.040]

Seeholzer, S, Tsuchimatsu, T, Jordan, T, Bieri, S, Pajonk, S, Yang, W, Jahoor, A, Shimizu, KK, Keller, B & Schulze-Lefert, P (2010) Diversity at the *Mla* Powdery Mildew Resistance Locus From Cultivated Barley Reveals Sites of Positive Selection. *Mol Plant Microbe Interact,* 23, 497-509.
[http://dx.doi.org/10.1094/MPMI-23-4-0497]

Segarra, G, Van der Ent, S, Trillas, I & Pieterse, CMJ (2009) MYB72, a node of convergence in induced systemic resistance triggered by a fungal and a bacterial beneficial Microbe. *Plant Biol,* 11, 90-6.
[http://dx.doi.org/10.1111/j.1438-8677.2008.00162.x]

Segura, A, Moreno, M, Molina, A & García-Olmedo, F (1998) Novel defensin subfamily from spinach (*Spinacia oleracea*). *FEBS Lett,* 435, 159-62.
[http://dx.doi.org/10.1016/S0014-5793(98)01060-6]

Segura, V, Vilhjálmsson, BJ, Platt, A, Korte, A, Seren, Ü, Long, Q & Nordborg, M (2012) An efficient multi-locus mixed model approach for genome-wide association studies in structured populations. *Nat Genet,* 44, 825-30.
[http://dx.doi.org/10.1038/ng.2314]

Sekhwal, MK, Li, P, Lam, I, Wang, X, Cloutier, S & You, FM (2015) Disease Resistance Gene Analogs (RGAs) in Plants. *Int J Mol Sci,* 16, 19248-90.
[http://dx.doi.org/10.3390/ijms160819248]

Selin, C, de Kievit, TR, Belmonte, MF & Fernando, WG (2016) Elucidating the Role of Effectors in Plant-Fungal Interactions: Progress and Challenges. *Front Microbiol,* 7, 600.
[http://dx.doi.org/10.3389/fmicb.2016.00600]

Selitrennikoff, CP (2001) Antifungal proteins. *Appl Environ Microbiol,* 67, 2883-94.
[http://dx.doi.org/10.1128/AEM.67.7.2883-2894.2001]

Sels, J, Mathys, J, De Coninck, BMA, Cammue, BPA & De Bolle, MFC (2008) Plant pathogenesis-related (PR) proteins: A focus on PR peptides. *Plant Physiol Biochem,* 46, 941-50.
[http://dx.doi.org/10.1016/j.plaphy.2008.06.011]

Senthil-Kumar, M & Mysore, KS (2013) Nonhost resistance against bacterial pathogens: retrospectives and prospects. *Annu Rev Phytopathol,* 51, 407-27.
[http://dx.doi.org/10.1146/annurev-phyto-082712-102319]

Shah, J (2003) The salicylic acid loop in plant defense. *Curr Opin Plant Biol,* 6, 365-71.
[http://dx.doi.org/10.1016/S1369-5266(03)00058-X]

Shah, J & Zeier, J (2013) Long-distance communication and signal amplification in systemic acquired resistance. *Front Plant Sci,* 4, 30.
[http://dx.doi.org/10.3389/fpls.2013.00030]

Shao, H, Wang, H & Tang, X (2015) NAC transcription factors in plant multiple abiotic stress responses: progress and prospects. *Front Plant Sci,* 6, 902.
[http://dx.doi.org/10.3389/fpls.2015.00902]

Shao, ZQ, Xue, JY, Wu, P, Zhang, YM, Wu, Y, Hang, YY, Wang, B & Chen, JQ (2016) Large-Scale Analyses of Angiosperm Nucleotide-Binding Site-Leucine-Rich Repeat Genes Reveal Three Anciently Diverged Classes with Distinct Evolutionary Patterns. *Plant Physiol,* 170, 2095-109.
[http://dx.doi.org/10.1104/pp.15.01487]

Shapiro, M, Salamouny, SE, Shepard, BM & Jackson, DM (2009) Plant Phenolics as Radiation Protectants for the Beet Armyworm (Lepidoptera: Noctuidae) Nucleopolyhedrovirus. *J Agric Urban Entomol,* 26, 1-10.
[http://dx.doi.org/10.3954/1523-5475-26.1.1]

Sharma, M & Pandey, GK (2015) Expansion and function of repeat domain proteins during stress and development in plants. *Front Plant Sci,* 6, 1218.
[http://dx.doi.org/10.3389/fpls.2015.01218]

Sharma, HC, Sharma, KK & Crouch, JH (2004) Genetic transformation of crops for insect resistance: potential and limitations. *Crit Rev Plant Sci,* 23, 47-72.
[http://dx.doi.org/10.1080/07352680490273400]

Shelton, AM & Badenes-Perez, FR (2006) Concepts and applications of trap cropping in pest management. *Annu Rev Entomol,* 51, 285-308.
[http://dx.doi.org/10.1146/annurev.ento.51.110104.150959]

Shen, Q, Liu, L, Wang, L & Wang, Q (2018) Indole primes plant defense against necrotrophic fungal pathogen infection. *PLoS One,* 13, e0207607.
[http://dx.doi.org/10.1371/journal.pone.0207607]

Shen, QH, Zhou, F, Bieri, S, Haizel, T, Shirasu, K & Schulze-Lefert, P (2003) Recognition specificity and RAR1/SGT1 dependence in barley Mla disease resistance genes to the powdery mildew fungus. *Plant Cell,* 15, 732-44.
[http://dx.doi.org/10.1105/tpc.009258]

Shibuya, K, Barry, KG, Ciardi, JA, Loucas, HM, Underwood, BA, Nourizadeh, S, Ecker, JR, Klee, HJ & Clark, DG (2004) The central role of PhEIN2 in ethylene responses throughout plant development in Petunia.

Plant Physiol, 136, 2900-12.
[http://dx.doi.org/10.1104/pp.104.046979]

Shigyo, M, Hasebe, M & Ito, M (2006) Molecular evolution of the AP2 subfamily. *Gene,* 366, 256-65.
[http://dx.doi.org/10.1016/j.gene.2005.08.009]

Shiu, SH & Bleecker, AB (2003) Expansion of the receptor-like kinase/Pelle gene family and receptor-like proteins in Arabidopsis. *Plant Physiol,* 132, 530-43.
[http://dx.doi.org/10.1104/pp.103.021964]

Silveira, AB, Trontin, C, Cortijo, S, Barau, J, Del Bem, LEV, Loudet, O, Colot, V & Vincentz, M (2013) Extensive natural epigenetic variation at a *de novo* originated gene. *PLoS Genet,* 9, e1003437.
[http://dx.doi.org/10.1371/journal.pgen.1003437]

Sinapidou, E, Williams, K, Nott, L, Bahkt, S, Tör, M, Crute, I, Bittner-Eddy, P & Beynon, J (2004) Two TIR:NB:LRR genes are required to specify resistance to *Peronospora parasitica* isolate Cala2 in *Arabidopsis.* *Plant J,* 38, 898-909.
[http://dx.doi.org/10.1111/j.1365-313X.2004.02099.x]

Singh, KB, Foley, RC & Oñate-Sánchez, L (2002) Transcription factors in plant defense and stress responses. *Curr Opin Plant Biol,* 5, 430-6.
[http://dx.doi.org/10.1016/s1369-5266(02)00289-3]

Sinha, AK, Jaggi, M, Raghuram, B & Tuteja, N (2011) Mitogen-activated protein kinase signaling in plants under abiotic stress. *Plant Signal Behav,* 6, 196-203.
[http://dx.doi.org/10.4161/psb.6.2.14701]

Sinha, M, Singh, RP, Kushwaha, GS, Iqbal, N, Singh, A, Kaushik, S, Kaur, P, Sharma, S & Singh, TP (2014) Current Overview of Allergens of Plant Pathogenesis Related Protein Families. *ScientificWorldJournal,* 2014, 1-19.
[http://dx.doi.org/10.1155/2014/543195]

Sinzelle, L, Izsvák, Z & Ivics, Z (2009) Molecular Domestication of Transposable Elements: From Detrimental Parasites to Useful Host Genes. *Cell Mol Life Sci,* 66, 1073-93.
[http://dx.doi.org/10.1007/s00018-009-8376-3]

Slaughter, A, Daniel, X, Flors, V, Luna, E, Hohn, B & Mauch-Mani, B (2012) Descendants of primed arabidopsis plants exhibit resistance to biotic stress. *Plant Physiol,* 158, 835-43.
[http://dx.doi.org/10.1104/pp.111.191593]

Snedden, WA & Fromm, H (2001) Calmodulin as a versatile calcium signal transducer in plants. *New Phytol,* 151, 35-66.
[http://dx.doi.org/10.1046/j.1469-8137.2001.00154.x]

Song, H, Guo, Z & Hu, X (2019) Evolutionary balance between LRR domain loss and young *NBS–LRR* genes production governs disease resistance in *Arachis hypogaea* cv. Tifrunner. *BMC Genomics,* 20, 844.
[http://dx.doi.org/10.1186/s12864-019-6212-1]

Song, H & Nan, Z (2014) Genome-wide analysis of nucleotide-binding site disease resistance genes in *Medicago truncatula. Chin Sci Bull,* 59, 1129-38.
[http://dx.doi.org/10.1007/s11434-014-0155-3]

Song, WY, Wang, GL, Chen, LL, Kim, HS, Pi, LY, Holsten, T, Gardner, J, Wang, B, Zhai, WX, Zhu, LH, Fauquet, C & Ronald, P (1995) A receptor kinase-like protein encoded by the rice disease resistance gene, *Xa21. Science,* 270, 1804-6.

[http://dx.doi.org/10.1126/science.270.5243.1804]

Souer, E, van Houwelingen, A, Kloos, D, Mol, J & Koes, R (1996) The no apical meristem gene of Petunia is required for pattern formation in embryos and flowers and is expressed at meristem and primordia boundaries. *Cell,*   85, 159-70.
[http://dx.doi.org/10.1016/S0092-8674(00)81093-4]

Soylu, S (2006) Accumulation of cell-wall bound phenolic compounds and phytoalexin in *Arabidopsis thaliana* leaves following inoculation with pathovars of *Pseudomonas syringae. Plant Sci,* 170, 942-52.
[http://dx.doi.org/10.1016/j.plantsci.2005.12.017]

Spielmeyer, W, Robertson, M, Collins, N, Leister, D, Schulze-Lefert, P, Seah, S, Moullet, O & Lagudah, ES (1998) A superfamily of disease resistance gene analogs is located on all homoeologous chromosome groups of wheat (Triticum aestivum). *Genome,* 41, 782-8.
[http://dx.doi.org/10.1139/g98-083]

Srichumpa, P, Brunner, S, Keller, B & Yahiaoui, N (2005) Allelic Series of Four Powdery Mildew Resistance Genes at the *Pm3* Locus in Hexaploid Bread Wheat. *Plant Physiol,* 139, 885-95.
[http://dx.doi.org/10.1104/pp.105.062406]

Stall, RE, Jones, JB & Minsavage, GV (2009) Durability of resistance in tomato and pepper to xanthomonads causing bacterial spot. *Annu Rev Phytopathol,* 47, 265-84.
[http://dx.doi.org/10.1146/annurev-phyto-080508-081752]

Staswick, PE & Tiryaki, I (2004) The oxylipin signal jasmonic acid is activated by an enzyme that conjugates it to isoleucine in Arabidopsis. *Plant Cell,* 16, 2117-27.
[http://dx.doi.org/10.1105/tpc.104.023549]

Stemple, DL (2004) TILLING - a high-throughput harvest for functional genomics. *Natl Rev,* 5, 145-50.
[http://dx.doi.org/10.1038/nrg1273]

Steuernagel, B, Witek, K, Krattinger, SG, Ramirez-Gonzalez, RH, Schoonbeek, HJ, Yu, G, Baggs, E, Witek, AI, Yadav, I, Krasileva, KV, Jones, JDG, Uauy, C, Keller, B & Ridout, CJ (2018) Physical and transcriptional organisation of the bread wheat intracellular immune receptor repertoire. bioRxiv. *Corpus ID,* 90166442
[http://dx.doi.org/10.1101/339424]

Stotz, HU, Thomson, JG & Wang, Y (2009) Plant defensins. Defense, development and application. *Plant Signal Behav,* 4, 1010-2.
[http://dx.doi.org/10.4161/psb.4.11.9755]

Straub, T, Ludewig, U & Neuhäuser, B (2017) The kinase CIPK23 inhibits ammonium transport in *Arabidopsis thaliana. Plant Cell,* 29, 409-22.
[http://dx.doi.org/10.1105/tpc.16.00806]

Stuthman, DD, Leonard, JJ & Miller-Garvin, J (2007) Breeding crops for durable resistance to disease. *Adv Agron,* 95, 319-67.
[http://dx.doi.org/10.1016/S0065-2113(07)95004-X]

Surridge, C (2015) Plant defence: Rubber bullets. *Nat Plants,* 1, 15026.
[http://dx.doi.org/10.1038/nplants.2015.26]

Swiderski, MR & Innes, RW (2001) The Arabidopsis PBS1 resistance gene encodes a member of a novel protein kinase subfamilly. *Plant J,* 26, 101-12.

[http://dx.doi.org/10.1046/j.1365-313x.2001.01014.x]

Szarejko, I & Szurman-Zubrzycka, M (2017) Creation of a TILLING Population in Barley After Chemical Mutagenesis with Sodium Azide and MNU.*Biotechnologies for Plant Mutation Breeding: Protocols,* Springer International Publishing 91-111.

Tabashnik, BE, Brévault, T & Carrière, Y (2013) Insect resistance to Bt crops: lessons from the first billion acres. *Nat Biotechnol,* 31, 510-21.
[http://dx.doi.org/10.1038/nbt.2597]

Tabashnik, BE, Mota-Sanchez, D, Whalon, ME, Hollingworth, RM & Carrière, Y (2014) Defining Terms for Proactive Management of Resistance to Bt Crops and Pesticides. *J Econ Entomol,* 107, 496-507.
[http://dx.doi.org/10.1603/EC13458]

Tabashnik, BE & Carrière, Y (2015) Successess and failures of transgenic Bt crops: Global patterns of field-evolved resistance.*Bt resistance: Characterization and strategies for GM crops producing Bacillus thuringiensis toxins,* CABI Press 1-14.

Takai, R, Isogai, A, Takayama, S & Che, FS (2008) Analysis of flagellin perception mediated by flg22 receptor OsFLS2 in rice. *Mol Plant Microbe Interact,* 21, 1635-42.
[http://dx.doi.org/10.1094/MPMI-21-12-1635]

Takeuchi, K, Taguchi, F, Inagaki, Y, Toyoda, K, Shiraishi, T & Ichinose, Y (2003) Flagellin glycosylation island in *Pseudomonas syringae* pv. *glycinea* and its role in host specificity. *J Bacteriol,* 185, 6658-65.
[http://dx.doi.org/10.1128/jb.185.22.6658-6665.2003]

Tan, L, Ijaz, U, Salih, H, Cheng, Z, Htet, NNW, Ge, Y & Azeem, F (2020) Genome-Wide Identification and Comparative Analysis of MYB Transcription Factor Family in *Musa acuminata* and *Musa balbisiana. Plants (Basel),* 9, E413.
[http://dx.doi.org/10.3390/plants9040413]

Tan, S & Wu, S (2012) Genome wide analysis of nucleotide-binding site disease resistance genes in *Brachypodium distachyon. Comp Funct Genomics,* 418208
[http://dx.doi.org/10.1155/2012/418208]

Tarr, DEK & Alexander, HM (2009) TIR-NBS-LRR genes are rare in monocots: evidence from diverse monocot orders. *BMC Res Notes,* 2, 197.
[http://dx.doi.org/10.1186/1756-0500-2-197]

Ten Have, A, Tenberge, KB, Benen, JAE, Tudzynski, P, Visser, J & van Kan, JAL (2002) The Contribution of Cell Wall Degrading Enzymes to Pathogenesis of Fungal Plant Pathogens.*Agricultural Applications The Mycota (A Comprehensive Treatise on Fungi as Experimental Systems for Basic and Applied Research),* Springer, Berlin, Heidelberg 341-58.

Terras, FR, Eggermont, K, Kovaleva, V, Raikhel, NV, Osborn, RW, Kester, A, Rees, SB, Torrekens, S, Van Leuven, F & Vanderleyden, J (1995) Small cysteine-rich antifungal proteins from radish: their role in host defense. *Plant Cell,* 7, 573-88.
[http://dx.doi.org/10.1105/tpc.7.5.573]

Terras, FR, Schoofs, HM, De Bolle, MF, Van Leuven, F, Rees, SB, Vanderleyden, J, Cammue, JP & Broekaert, WF (1992) Analysis of Two Novel Classes of Plant Antifungal Proteins From Radish (*Raphanus Sativus* L.) Seeds. *J Biol Chem,* 267, 15301-9.

Thabuis, A, Lefebvre, V, Bernard, G, Daubèze, AM, Phaly, T, Pochard, E & Palloix, A (2004) Phenotypic and molecular evaluation of a recurrent selection program for a polygenic resistance to *Phytophthora capsici*

in pepper. *Theor Appl Genet,* 109, 342-51.
[http://dx.doi.org/10.1007/s00122-004-1633-9]

Theis, T & Stahl, U (2004) Antifungal proteins: targets, mechanisms and prospective applications. *Cell Mol Life Sci,* 61, 437-55.
[http://dx.doi.org/10.1007/s00018-003-3231-4]

Thines, B, Katsir, L, Melotto, M, Niu, Y, Mandaokar, A, Liu, GH, Nomura, K, He, SY, Howe, GA & Browse, J (2007) JAZ repressor proteins are targets of the SCFCO11 complex during jasmonate signalling. *Nature,* 448, 661-2.
[http://dx.doi.org/10.1038/nature05960]

Thines, M & Kamoun, S (2010) Oomycete-plant Coevolution: Recent Advances and Future Prospects. *Curr Opin Plant Biol,* 13, 427-33.
[http://dx.doi.org/10.1016/j.pbi.2010.04.001]

Thomas, CM, Jones, DA, Parniske, M, Harrison, K, Balint-Kurti, PJ, Hatzixanthis, K & Jones, JD (1997) Characterization of the Tomato Cf-4 Gene for Resistance to *Cladosporium Fulvum* Identifies Sequences That Determine Recognitional Specificity in Cf-4 and Cf-9. *Plant Cell,* 9, 2209-24.
[http://dx.doi.org/10.1105/tpc.9.12.2209]

Thomma, BPHJ, Eggermont, K, Penninckx, IAMA, Mauch-Mani, B, Vogelsang, R, Cammue, BPA & Broekaert, WF (1998) Separate jasmonate-dependent and salicylate-dependent defense response pathways in *Arabidopsis* are essential for resistance to distinct microbial pathogens. *Proc Natl Acad Sci USA,* 95, 15107-11.
[http://dx.doi.org/10.1073/pnas.95.25.15107]

Thomzik, JE, Stenzel, K, Stöcker, R, Schreier, PH, Hain, R & Stahl, DJ (1997) Synthesis of a grapevine phytoalexin in transgenic tomatoes (*Lycopersicon esculentum* Mill.) conditions resistance against *Phytophthora infestans. Physiol Mol Plant Pathol,* 51, 265-78.
[http://dx.doi.org/10.1006/pmpp.1997.0123]

Thynne, E, McDonald, MC & Solomon, PS (2015) Phytopathogen emergence in the genomics era. *Trends Plant Sci,* 20, 246-55.
[http://dx.doi.org/10.1016/j.tplants.2015.01.009]

Tian, D & Yin, Z (2009) Constitutive heterologous expression of *avrXa27* in rice containing the R gene *Xa27* confers enhanced resistance to compatible *Xanthomonas oryzae* strains. *Mol Plant Pathol,* 10, 29-39.
[http://dx.doi.org/10.1111/j.1364-3703.2008.00509.x]

Till, BJ, Reynolds, SH, Greene, EA, Codomo, CA, Enns, LC, Johnson, JE, Burtner, C, Odden, AR, Young, K, Taylor, NE, Henikoff, JG, Comai, L & Henikoff, S (2003) Large-scale discovery of induced point mutations with high-throughput TILLING. *Genome Res,* 13, 524-30.
[http://dx.doi.org/10.1101/gr.977903]

Richter, TE & Ronald, PC (2000) The evolution of disease resistance genes. *Plant Mol Biol,* 42, 195-204.
[http://dx.doi.org/10.1023/A:1006388223475]

Palmer, TM, Stanton, ML, Young, TP, Goheen, JR, Pringle, RM & Karban, R (2008) Breakdown of an ant-plant mutualism follows the loss of large herbivores from an African savanna. *Science,* 319, 192-5.
[http://dx.doi.org/10.1126/science.1151579]

Tör, M, Lotze, MT & Holton, N (2009) Receptor-mediated signalling in plants: molecular patterns and programmes. *J Exp Bot,* 60, 3645-54.

[http://dx.doi.org/10.1093/jxb/erp233]

Torres, MA & Dangl, JL (2005) Functions of the respiratory burst oxidase in biotic interactions, abiotic stress and development. *Curr Opin Plant Biol,* 8, 397-403.
[http://dx.doi.org/10.1016/j.pbi.2005.05.014]

Torres, MA, Jones, JDG & Dangl, JL (2006) Reactive Oxygen Species Signaling in Response to Pathogens. *Plant Physiol,* 141, 373-8.
[http://dx.doi.org/10.1104/pp.106.079467]

Tran, LS, Nakashima, K, Sakuma, Y, Simpson, SD, Fujita, Y, Maruyama, K, Fujita, M, Seki, M, Shinozaki, K & Yamaguchi-Shinozaki, K (2004) Isolation and functional analysis of *Arabidopsis* stress-inducible NAC transcription factors that bind to a drought-responsive cis-element in the early responsive to dehydration stress 1 promoter. *Plant Cell,* 16, 2481-98.
[http://dx.doi.org/10.1105/tpc.104.022699]

Tratwal, A & Bocianowski, J (2018) Cultivar mixtures as part of integrated protection of spring barley. *J Plant Dis Prot,* 125, 41-50.
[http://dx.doi.org/10.1007/s41348-017-0139-z]

Trdá, L, Fernandez, O, Boutrot, F, Héloir, MC, Kelloniemi, J, Daire, X, Adrian, M, Clément, C, Zipfel, C, Dorey, S & Poinssot, B (2014) The grapevine flagellin receptor VvFLS2 differentially recognizes flagellin-derived epitopes from the endophytic growth promoting bacterium *Burkholderia phytofirmans* and plant pathogenic bacteria. *New Phytol,* 201, 1371-84.
[http://dx.doi.org/10.1111/nph.12592]

Treutter, D (2000) Induced resistance in plant pathology, consequences for the quality of plant foodstuffs? *J Appl Bot,* 74, 1-4.

Tripathi, JN, Lorenzen, J, Bahar, O, Ronald, P & Tripathi, L (2014) Transgenic expression of the rice *Xa21* pattern-recognition receptor in banana (*Musa* sp.) confers resistance to *Xanthomonas campestris* pv. *musacearum. Plant Biotechnol J,* 12, 663-73.
[http://dx.doi.org/10.1111/pbi.12170]

Triques, K, Sturbois, B, Gallais, S, Dalmais, M, Chauvin, S, Clepet, C, Aubourg, S, Rameau, C, Caboche, M & Bendahmane, A (2007) Characterization of *Arabidopsis thaliana* mismatch specific endonucleases Application to mutation discovery by TILLING in pea. *Plant J,* 51, 1116-25.
[http://dx.doi.org/10.1111/j.1365-313X.2007.03201.x]

Trowsdale, J (2002) The gentle art of gene arrangement: the meaning of gene clusters. *Genome Biology,* 3, 2002.1-5.

Tsuda, K, Sato, M, Stoddard, T, Glazebrook, J & Katagiri, F (2009) Network properties of robust immunity in plants. *PLoS Genet,* 5, e1000772.
[http://dx.doi.org/10.1371/journal.pgen.1000772]

Tuteja, N & Mahajan, S (2007) Calcium signaling network in plants: an overview. *Plant Signal Behav,* 2, 79-85.
[http://dx.doi.org/10.4161/psb.2.2.4176]

Vallad, GE & Goodman, RM (2004) Systemic acquired resistance and induced systemic resistance in conventional agriculture. *Crop Sci,* 44, 1920.
[http://dx.doi.org/10.2135/cropsci2004.1920]

Van Damme, M, Andel, A, Huibers, RP, Panstruga, R, Weisbeek, PJ & Van den Ackerveken, G (2005)

Identification of *Arabidopsis* loci required for susceptibility to the downy mildew pathogen *Hyaloperonospora parasitica. Mol Plant Microbe Interact,* 18, 583-92.
[http://dx.doi.org/10.1094/MPMI-18-0583]

Van Damme, M, Huibers, RP, Elberse, J & Van den Ackerveken, G (2008) *Arabidopsis* DMR6 encodes a putative 2OG-Fe(II) oxygenase that is defense-associated but required for susceptibility to downy mildew. *Plant J,* 54, 785-93.
[http://dx.doi.org/10.1111/j.1365-313X.2008.03427.x]

van den Bosch, F & Gilligan, CA (2003) Measures of Durability of Resistance. *Phytopathology,* 93, 616-25.
[http://dx.doi.org/10.1094/PHYTO.2003.93.5.616]

Van Der Biezen, EA & Jones, JDG (1998) The NB-ARC domain: a novel signalling motif shared by plant resistance gene products and regulators of cell death in animals. *Curr Biol,* 8, R226-7. a
[http://dx.doi.org/10.1016/s0960-9822(98)70145-9]

Van Der Biezen, EA & Jones, JDG (1998) Plant disease-resistance proteins and the gene-for gene concept. *Trends Biochem Sci,* 23, 454-6. b
[http://dx.doi.org/10.1016/S0968-0004(98)01311-5]

Van der Ent, S, Verhagen, BW, Van Doorn, R, Bakker, D, Verlaan, MG, Pel, MJ, Joosten, RG, Proveniers, MC, Van Loon, LC, Ton, J & Pieterse, CM (2008) MYB72 is required in early signaling steps of rhizobacteria-induced systemic resistance in *Arabidopsis. Plant Physiol,* 146, 1293-304.
[http://dx.doi.org/10.1104/pp.107.113829]

Van der Hoorn, RAL & Kamoun, S (2008) From Guard to Decoy: A New Model for Perception of Plant Pathogen Effectors. *Plant Cell,* 20, 2009-17.
[http://dx.doi.org/10.1105/tpc.108.060194]

Van Loon, LC (1999) Occurrence and properties of plant pathogenesis-related proteins. In: Datta, SK, Muthukrishnan, S, (Eds.), *Pathogenesis-related proteins in plants,* CRC Press, Roca Raton 1-19.

Van Loon, LC, Rep, M & Pieterse, CM (2006) Significance of inducible defense-related proteins in infected plants. *Annu Rev Phytopathol,* 44, 135-62.
[http://dx.doi.org/10.1146/annurev.phyto.44.070505.143425]

Van Loon, LC & Van Kammen, A (1970) Polyacrylamide disc electrophoresis of the soluble leaf proteins from *Nicotiana tabacum* var. *samsun* and *samsun NN*: II. Changes in protein constitution after infection with tobacco mosaic virus. *Virology,* 40, 199-211.
[http://dx.doi.org/10.1016/0042-6822(70)90395-8]

van Ooijen, G, van den Burg, HA, Cornelissen, BJC & Takken, FLW (2007) Structure and Function of Resistance Proteins in Solanaceous Plants. *Annu Rev Phytopathol,* 45, 43-72.
[http://dx.doi.org/10.1146/annurev.phyto.45.062806.094430]

Van Schie, CCN & Takken, FLW (2014) Susceptibility Genes 101: How to Be a Good Host. *Annu Rev Phytopathol,* 52, 551-81.
[http://dx.doi.org/10.1146/annurev-phyto-102313-045854]

Velculescu, VE, Zhang, L, Vogelstein, B & Kinzler, KW (1995) Serial analysis of gene expression. *Science,* 270, 484-7.
[http://dx.doi.org/10.1126/science.270.5235.484]

Vemanna, RS, Bakade, R, Bharti, P, Kumar, MKP, Sreeman, SM, Senthil-Kumar, M & Makarla, U (2019) Cross-talk signaling in rice during combined drought and bacterial blight stress. *Front Plant Sci,* 10, 193.

[http://dx.doi.org/10.3389/fpls.2019.00193]

Venu, RC, Jia, Y, Gowda, M, Jia, MH & Jantasuriyarat, C (2007) RL-SAGE and Microarray Analysis of the Rice Transcriptome After *Rhizoctonia Solani* Infection. *Mol Genet Genomics,* 278, 421-31.
[http://dx.doi.org/10.1007/s00438-007-0260-y]

Verma, S, Gazara, RK & Verma, PK (2017) Transcription Factor Repertoire of Necrotrophic Fungal Phytopathogen *Ascochyta rabiei*: Predominance of MYB Transcription Factors As Potential Regulators of Secretome. *Front Plant Sci,* 8, 1037.
[http://dx.doi.org/10.3389/fpls.2017.01037]

Vermerris, W & Nicholson, R (2008) The Role of Phenols in Plant Defense.*Phenolic Compound Biochemistry,* Springer, Dordrecht 211-34.

Vetter, MM, Kronholm, I, He, F & Häweker, H (2012) Flagellin Perception Varies Quantitatively in *Arabidopsis thaliana* and Its Relatives. *Mol Biol Evol,* 29, 1655-67.
[http://dx.doi.org/10.1093/molbev/mss011]

Visser, B, Le Lann, C, Den Blanken, FJ, Harvey, JA, Van Alphen, JJ & Ellers, J (2010) Loss of lipid synthesis as an evolutionary consequence of a parasitic lifestyle. *Proc Natl Acad Sci USA,* 107, 8677-82.
[http://dx.doi.org/10.1073/pnas.1001744107]

Vitte, C & Bennetzen, JL (2006) Analysis of retrotransposon structural diversity uncovers properties and propensities in angiosperm genome evolution. *Proc Natl Acad Sci USA,* 103, 17638-43.
[http://dx.doi.org/10.1073/pnas.0605618103]

Vleesschauwer, D & Hofte, M (2009) Rhizobacteria-induced systemic resistance.*Plant Innate Immunity,* Academic Press Ltd-Elsevier Science Ltd, London Vol. 51, 223-81.

Vlot, AC, Dempsey, DA & Klessig, DF (2009) Salicylic acid, a multifaceted hormone to combat disease. *Annu Rev Phytopathol,* 47, 177-206.
[http://dx.doi.org/10.1146/annurev.phyto.050908.135202]

Vos, P, Hogers, R, Bleeker, M, Reijans, M, van de Lee, T, Hornes, M, Frijters, A, Pot, J, Peleman, J & Kuiper, M (1995) AFLP: A New Technique for DNA Fingerprinting. *Nucleic Acids Res,* 23, 4407-14.
[http://dx.doi.org/10.1093/nar/23.21.4407]

Wada, Y, Miyamoto, K, Kusano, T & Sano, H (2004) Association between up-regulation of stress-responsive genes and hypomethylation of genomic DNA in tobacco plants. *Mol Genet Genomics,* 271, 658-66.
[http://dx.doi.org/10.1007/s00438-004-1018-4]

Wan, H, Yuan, W, Bo, K & Shen, J (2013) Genome-wide analysis of NBS-encoding disease resistance genes in *Cucumis sativus* and phylogenetic study of NBS-encoding genes in Cucurbitaceae crops. *BMC Genomics,* 14, 109.
[http://dx.doi.org/10.1186/1471-2164-14-109]

Wang, B, Yu, J, Zhu, D & Zhao, Q (2011) Maize defensin ZmDEF1 is involved in plant response to fungal phytopathogens. *Afr J Biotechnol,* 10, 16128-37.
[http://dx.doi.org/10.5897/AJB11.1456]

Wang, C-IA, Guncar, G & Forwood, JK (2007) Crystal Structures of Flax Rust Avirulence Proteins AvrL567-A and -D Reveal Details of the Structural Basis for Flax Disease Resistance Specificity. *Plant Cell,* 19, 2898-912.

[http://dx.doi.org/10.1105/tpc.107.053611]

Wang, F, Lin, R, Feng, J, Chen, W, Qiu, D & Xu, S (2015) TaNAC1 acts as a negative regulator of stripe rust resistance in wheat, enhances susceptibility to *Pseudomonas syringae*, and promotes lateral root development in transgenic *Arabidopsis thaliana. Front Plant Sci,* 6, 108.
[http://dx.doi.org/10.3389/fpls.2015.00108]

Wang, G, Ellendorff, U, Kemp, B, Mansfield, JW, Forsyth, A, Mitchell, K, Bastas, K, Liu, C-M, Woods-Tör, A, Zipfel, C, de Wit, PJGM, Jones, JDG, Tör, M & Thomma, BPHJ (2008) A Genome-Wide Functional Investigation into the Roles of Receptor-Like Proteins in Arabidopsis. *Plant Physiol,* 147, 503-17.
[http://dx.doi.org/10.1104/pp.108.119487]

Wang, H, Wang, H, Shao, H & Tang, X (2016) Recent Advances in Utilizing Transcription Factors to Improve Plant Abiotic Stress Tolerance by Transgenic Technology. *Front Plant Sci,* 7, 1-13. a
[http://dx.doi.org/10.3389/fpls.2016.00067]

Wang, JF, Hanson, P & Barnes, JA (1998) Worldwide evaluation of an international set of resistant sources to bacterial wilt in tomato.*Bacterial Wilt Disease: Molecular and Ecological Aspects,* Springer Verlag 269-75.

Wang, Y, Cheng, X, Shan, Q, Zhang, Y, Liu, J, Gao, C & Qiu, J-L (2014) Simultaneous Editing of Three Homoeoalleles in Hexaploid Bread Wheat Confers Heritable Resistance to Powdery Mildew (2014). *Nat Biotechnol,* 32, 947-51.
[http://dx.doi.org/10.1038/nbt.2969]

Wang, Y, Zhang, Y, Ji, W, Yu, P, Wang, B, Li, J, Han, M, Xu, X & Wang, Z (2016) Cultivar Mixture Cropping Increased Water Use Efficiency in Winter Wheat under Limited Irrigation Conditions. *PLoS One,* 11, e0158439. b
[http://dx.doi.org/10.1371/journal.pone.0158439]

Wang, Z, Gerstein, M & Snyder, M (2009) RNA-seq: a revolutionary tool for transcriptomics. *Nat Rev Genet,* 10, 57-63.
[http://dx.doi.org/10.1038/nrg2484]

Waqas, M, Azhar, MT, Rana, IA, Azeem, F, Ali, MA, Nawaz, MA, Chuang, G & Atif, RM (2019) Genome-wide identification and expression analyses of WRKY transcription factor family members from chickpea (*Cicer arietinum* L.) reveal their role in abiotic stress-responses. *Genes Genomics,* 41, 467-81.
[http://dx.doi.org/10.1007/s13258-018-00780-9]

Warren, RF, Henk, A, Mowery, P, Holub, E & Innes, RW (1998) A mutation within the leucine-rich repeat domain of the *Arabidopsis* disease resistance gene *RPS5* partially suppresses multiple bacterial and downy mildew resistance genes. *Plant Cell,* 10, 1439-52.
[http://dx.doi.org/10.1105/tpc.10.9.1439]

Wei, F, Gobelman-Werner, K, Morroll, SM, Kurth, J, Mao, L, Wing, R, Leister, D, Schulze-Lefert, P & Wise, RP (1999) The Mla (powdery mildew) resistance cluster is associated with three NBS-LRR gene families and suppressed recombination within a 240-kb DNA interval on chromosome 5S (1HS) of barley. *Genetics,* 153, 1929-48. [Erratum in: Genetics 154, 953].

Wei, F, Wing, RA & Wise, RP (2002) Genome Dynamics and Evolution of the Mla (Powdery Mildew) Resistance Locus in Barley. *Plant Cell,* 14, 1903-17.
[http://dx.doi.org/10.1105/tpc.002238]

Wei, K, Chen, J, Wang, Y, Chen, Y, Chen, S, Lin, Y, Pan, S, Zhong, X & Xie, D (2012) Genome-wide analysis of bZIP-encoding genes in maize. *DNA Res,* 19, 463-76. a

[http://dx.doi.org/10.1093/dnares/dss026]

Wei, KF, Chen, J, Chen, YF, Wu, LJ & Xie, DX (2012) Molecular phylogenetic and expression analysis of the complete WRKY transcription factor family in maize. *DNA Res,* 19, 153-64. b
[http://dx.doi.org/10.1093/dnares/dsr048]

Wen, L, Chang, HX, Brown, PJ, Domier, LL & Hartman, GL (2019) Genome-wide association and genomic prediction identifies soybean cyst nematode resistance in common bean including a syntenic region to soybean Rhg1 locus. *Hortic Res,* 6, 9.
[http://dx.doi.org/10.1038/s41438-018-0085-3]

Whitham, S, Dinesh-Kumar, SP, Choi, D, Hehl, R, Corr, C & Baker, B (1994) The product of the tobacco mosaic virus resistance gene *N*: similarity to toll and the interleukin-1 receptor. *Cell,* 78, 1101-15.
[http://dx.doi.org/10.1016/0092-8674(94)90283-6]

Wiesner-Hanks, T & Nelson, R (2016) Multiple disease resistance in plants. *Annu Rev Phytopathol,* 54, 229-52.
[http://dx.doi.org/10.1146/annurev-phyto-080615-100037]

Wilcox, JR & St Martin, SK (1998) Soybean genotypes resistant to *Phytophthora sojae* and compensation for yield losses of susceptible isolines. *Plant Dis,* 82, 303-6.
[http://dx.doi.org/10.1094/PDIS.1998.82.3.303]

Wilde, F, Schön, CC, Korzun, V, Ebmeyer, E, Schmolke, M, Hartl, L & Miedaner, T (2008) Marker-based introduction of three quantitative-trait loci conferring resistance to Fusarium head blight into an independent elite winter wheat breeding population. *Theor Appl Genet,* 117, 29-35.
[http://dx.doi.org/10.1007/s00122-008-0749-8]

Wildermuth, MC, Dewdney, J, Wu, G & Ausubel, FM (2001) Isochorismate synthase is required to synthesize salicylic acid for plant defence. *Nature,* 414, 562-5.
[http://dx.doi.org/10.1038/35107108]

Williams, JGK, Kubelik, AR, Livak, KJ, Rafalski, JA & Tingey, SV (1990) DNA polymorphisms amplified by arbitrary primers are useful as genetic markers. *Nucleic Acids Res,* 18, 6531-5.
[http://dx.doi.org/10.1093/nar/18.22.6531]

Williams, SJ, Sohn, KH, Wan, L, Bernoux, M, Sarris, PF, Segonzac, C, Ve, T, Ma, Y, Saucet, SB & Ericsson, DJ (2014) Structural Basis for Assembly and Function of a Heterodimeric Plant Immune Receptor. *Science,* 344, 299-303.
[http://dx.doi.org/10.1126/science.1247357]

Wilson, CR (2014) *Applied plant virology,* CABI, UK.

Wilson, CR, Sauer, JM & Hooser, SB (2001) Taxines: a review of the mechanism and toxicity of yew (*Taxus* spp.) alkaloids. *Toxicon,* 39, 175-85.
[http://dx.doi.org/10.1016/S0041-0101(00)00146-X]

Wilton, M, Subramaniam, R, Elmore, J, Felsensteiner, C, Coaker, G & Desveaux, D (2010) The type III effector HopF2Pto targets Arabidopsis RIN4 protein to promote *Pseudomonas syringae* virulence. *Proc Natl Acad Sci USA,* 107, 2349-54.
[http://dx.doi.org/10.1073/pnas.0904739107]

Witsenboer, H, Kesseli, RV, Fortin, M, Stanghellini, M & Michelmore, RW (1995) Sources and genetic structure of a cluster of genes for resistance to three pathogens in lettuce. *Theor Appl Genet,* 91, 178-88.

Wójcicka, A (2015) Surface waxes as a plant defense barrier towards grain aphid. *Acta Biol Cracov Ser; Bot,* 57, 95-103.
[http://dx.doi.org/10.1515/abcsb-2015-0012]

Wojtaszek, P (1997) Oxidative burst: an early plant response to pathogen infection. *Biochem J,* 322, 681-92.
[http://dx.doi.org/10.1042/bj3220681]

Wong, JH, Xia, L & Ng, TB (2007) A review of defensins of diverse origins. *Curr Protein Pept Sci,* 8, 446-59.
[http://dx.doi.org/10.2174/138920307782411446]

Woolhouse, ME & Gowtage-Sequeria, S (2005) Host range and emerging and reemerging pathogens. *Emerg Infect Dis,* 11, 1842-7.
[http://dx.doi.org/10.3201/eid1112.050997]

Woolhouse, ME, Taylor, LH & Haydon, DT (2001) Population biology of multihost pathogens. *Science,* 292, 1109-12.
[http://dx.doi.org/10.1126/science.1059026]

Wouters, FC, Blanchette, B & Gershenzon, J (2016) Plant defense and herbivore counter-defense: benzoxazinoids and insect herbivores. *Phytochem Rev,* 15, 1127-51.
[http://dx.doi.org/10.1007/s11101-016-9481-1]

Wu, CH, Abd-El-Haliem, A, Bozkurt, TO, Belhaj, K, Terauchi, R, Vossen, JH & Kamoun, S (2017) NLR network mediates immunity to diverse plant pathogens. *Proc Natl Acad Sci USA,* 114, 8113-8.
[http://dx.doi.org/10.1073/pnas.1702041114]

Wu, C-H & Kamoun, S (2019) Tomato Prf requires NLR helpers NRC2 and NRC3 to confer resistance against the bacterial speck pathogen *Pseudomonas syringae* pv. *tomato. bioRxiv,* 595744
[http://dx.doi.org/10.1101/595744]

Wu, Q & VanEtten, HD (2004) Introduction of Plant and Fungal Genes Into Pea (Pisum Sativum L.) Hairy Roots Reduces Their Ability to Produce Pisatin and Affects Their Response to a Fungal Pathogen. *Mol Plant Microbe Interact,* 17, 798-804.
[http://dx.doi.org/10.1094/MPMI.2004.17.7.798]

Wu, Q, Wang, X & Ding, S-W (2010) Viral suppressors of RNA-based viral immunity: Host targets. *Cell Host Microbe,* 8, 12-5.
[http://dx.doi.org/10.1016/j.chom.2010.06.009]

Wu, Y & Zhou, JM (2013) Receptor-like kinases in plant innate immunity. *J Integr Plant Biol,* 55, 1271-86.
[http://dx.doi.org/10.1111/jipb.12123]

Xia, N, Zhang, G, Liu, X-Y, Deng, L, Cai, G-L, Zhang, Y, Wang, X-J, Zhao, J, Huang, L-L & Kang, Z-S (2010) Characterization of a Novel Wheat NAC Transcription Factor Gene Involved in Defense Response Against Stripe Rust Pathogen Infection and Abiotic Stresses. *Mol Biol Rep,* 37, 3703-12. b
[http://dx.doi.org/10.1007/s11033-010-0023-4]

Xia, N, Zhang, G, Sun, Y-F, Zhu, L, Xu, L-S, Chen, X-M, Liu, B, Yu, Y-T, Wang, X-J, Huang, L-L & Kang, Z-S (2010) TaNAC8, a novel NAC transcription factor gene in wheat, responds to stripe rust pathogen infection and abiotic stresses. *Physiol Mol Plant Pathol,* 74, 394-402. a
[http://dx.doi.org/10.1016/j.pmpp.2010.06.005]

Xiao, S, Ellwood, S, Calis, O, Patrick, E, Li, T, Coleman, M & Turner, JG (2001) Broad spectrum mildew

resistance in Arabidopsis thaliana mediated by RPW8. *Science,* 291, 118-20.
[http://dx.doi.org/10.1099/vir.0.82377-0]

Xu, C, Liu, R, Zhang, Q, Chen, X, Qian, Y & Fang, W (2017) The diversification of evolutionarily conserved MAPK cascades correlates with the evolution of fungal species and development of lifestyles. *Genome Biol Evol,* 9, 311-22.
[http://dx.doi.org/10.1093/gbe/evw051]

Xu, X, Chen, C, Fan, B & Chen, Z (2006) Physical and functional interactions between pathogen-induced Arabidopsis WRKY18, WRKY40, and WRKY60 transcription factors. *Plant Cell,* 18, 1310-26.
[http://dx.doi.org/10.1105/tpc.105.037523]

Xue, JY, Zhao, T, Liu, Y, Liu, Y, Zhang, Y-X, Zhang, G-Q, Chen, H, Zhou, G-C, Zhang, S-Z & Shao, Z-Q (2020) Genome-Wide Analysis of the Nucleotide Binding Site Leucine-Rich Repeat Genes of Four Orchids Revealed Extremely Low Numbers of Disease Resistance Genes. *Front Genet,* 10, 1286.
[http://dx.doi.org/10.3389/fgene.2019.01286]

Xue, JY, Wang, Y, Wu, P, Wang, Q, Yang, L-T, Pan, X-H, Wang, B & Chen, J-Q (2012) A primary survey on bryophyte species reveals two novel classes of nucleotide-binding site (NBS) genes. *PLoS One,* 7, e36700.
[http://dx.doi.org/10.1371/journal.pone.0036700]

Yalpani, N, Shulaev, VI & Raskin, I (1993) Endogenous salicylic acid levels correlate with accumulation of pathogenesis-related proteins and virus resistance in tobacco. *Phytopathology,* 83, 702-8.
[http://dx.doi.org/10.1094/Phyto-83-702]

Yamaguchi, Y, Pearce, G & Ryan, CA (2006) The cell surface leucine-rich repeat receptor for AtPep1, an endogenous peptide elicitor in Arabidopsis, is functional in transgenic tobacco cells. *Proc Natl Acad Sci USA,* 103, 10104-9.
[http://dx.doi.org/10.1073/pnas.0603729103]

Yang, S, Zhang, X, Yue, JX, Tian, D & Chen, JQ (2008) Recent duplications dominate NBS-encoding gene expansion in two woody species. *Mol Genet Genomics,* 280, 187-98.
[http://dx.doi.org/10.1007/s00438-008-0355-0]

Yang, Y, Li, J, Li, H, Yang, Y, Guang, Y & Zhou, Y (2019) The bZIP gene family in watermelon: genome-wide identification and expression analysis under cold stress and root-knot nematode infection. *PeerJ,* 7, e7878.
[http://dx.doi.org/10.7717/peerj.7878]

Yegutkin, GG (2008) Nucleotide- and nucleoside-converting ectoenzymes: Important modulators of purinergic signalling cascade. *Biochim Biophys Acta,* 1783, 673-94.
[http://dx.doi.org/10.1016/j.bbamcr.2008.01.024]

Yogendra, KN, Kushalappa, AC, Sarmiento, F, Rodriguez, E & Mosquera, T (2015) Metabolomics deciphers quantitative resistance mechanisms in diploid potato clones against late blight. *Funct Plant Biol,* 42, 284-98.
[http://dx.doi.org/10.1071/fp14177]

Yoshida, K, Saunders, DG, Mitsuoka, C, Natsume, S, Kosugi, S, Saitoh, H, Inoue, Y, Chuma, I, Tosa, Y & Cano, LM (2016) Host specialization of the blast fungus *Magnaporthe oryzae* is associated with dynamic gain and loss of genes linked to transposable elements. *BMC Genomics,* 17, 370.
[http://dx.doi.org/10.1186/s12864-016-2690-6]

Young, ND (2000) The genetic architecture of resistance. *Curr Opin Plant Biol,* 3, 285-90.

[http://dx.doi.org/10.1016/s1369-5266(00)00081-9]

Yu, A, Lepère, G, Jay, F, Wang, J, Bapaume, L, Wang, Y, Abraham, AL, Penterman, J, Fischer, RL & Voinnet, O (2013) Dynamics and biological relevance of DNA demethylation in Arabidopsis antibacterial defense. *Proc Natl Acad Sci USA,* 110, 2389-94.
[http://dx.doi.org/10.1073/pnas.1211757110]

Yu, J, Tehrim, S, Zhang, F, Tong, C & Huang, J (2014) Genome-wide comparative analysis of NBS-encoding genes between *Brassica* species and *Arabidopsis thaliana. BMC Genomics,* 15, 3.
[http://dx.doi.org/10.1186/1471-2164-15-3]

Yu, JM & Buckler, ES (2006) Genetic association mapping and genome organization of maize. *Current Opinion Biotechnology,* 17, 155-60.

Yue, J-X, Meyers, BC, Chen, J-Q, Tian, D & Yang, S (2012) Tracing the origin and evolutionary history of plant nucleotide-binding site-leucine-rich repeat (NBS-LRR) genes. *New Phytol,* 193, 1049-63.
[http://dx.doi.org/10.1111/j.1469-8137.2011.04006.x]

Yuen, B, Bayes, JM & Degnan, SM (2014) The characterization of sponge NLRs provides insight into the origin and evolution of this innate immune gene family in animals. *Mol Biol Evol,* 31, 106-20.
[http://dx.doi.org/10.1093/molbev/mst174]

Zanke, BW, Boudreau, K, Rubie, E, Winnett, E, Tibbles, LA & Zon, L (1996) The stress-activated protein kinase pathway mediates cell death following injury induced by cis-platinum, UV irradiation or heat. *Curr Biol,* 6, 606-13.
[http://dx.doi.org/10.1016/S0960-9822(02)00547-X]

Zeilmaker, T, Ludwig, NR, Elberse, J, Seidl, MF, Berke, L, Van Doorn, A, Schuurink, RC, Snel, B & Van den Ackerveken, G (2015) DOWNY MILDEW RESISTANT 6 and DMR6-LIKE OXYGENASE 1 are partially redundant but distinct suppressors of immunity in Arabidopsis. *Plant J,* 81, 210-22.
[http://dx.doi.org/10.1111/tpj.12719]

Zemach, A, McDaniel, IE, Silva, P & Zilberman, D (2010) Genome-wide evolutionary analysis of eukaryotic DNA methylation. *Science,* 328, 916-9.
[http://dx.doi.org/10.1126/science.1186366]

Zenbayashi-Sawata, K, Fukuoka, S, Katagiri, S, Fujisawa, M, Matsumoto, T, Ashizawa, T & Koizumi, S (2007) Genetic and Physical Mapping of the Partial Resistance Gene, pi34, to Blast in Rice. *Phytopathology,* 97, 598-602.
[http://dx.doi.org/10.1094/PHYTO-97-5-0598]

Zeng, X, Tian, D, Gu, K, Zhou, Z, Yang, X, Luo, Y, White, FF & Yin, Z (2015) Genetic engineering of the Xa10 promoter for broad-spectrum and durable resistance to *Xanthomonas oryzae* pv.*oryzae. Plant Biotechnol J,* 13, 993-1001.
[http://dx.doi.org/10.1111/pbi.12342]

Zernova, OV, Lygin, AV, Pawlowski, ML, Hill, CB, Hartman, GL, Widholm, JM & Lozovaya, VV (2014) Regulation of plant immunity through modulation of phytoalexin synthesis. *Molecules,* 19, 7480-96.

Zhang, T & Miao, J (2018) Na Han, Yujun Qiang, Wen Zhang, MPD (2018) MPD: a pathogen genome and metagenome database. *Database (Oxford),* 2018, bay055.
[http://dx.doi.org/10.1093/database/bay055]

Zhang, H & Sonnewald, U (2017) Differences and commonalities of plant responses to single and combined stresses. *Plant J,* 90, 839-55.

[http://dx.doi.org/10.1111/tpj.13557]

Zhang, X, Du, P & Lu, L (2008) Contrasting effects of HC-Pro and 2b viral suppressors from Sugarcane mosaic virus and Tomato aspermy cucumovirus on the accumulation of siRNAs. *Virology,* 374, 351-60.
[http://dx.doi.org/10.1016/j.virol.2007.12.045]

Zhang, X, Liang, P & Ming, R (2016) Genome-wide identification and characterization of nucleotide-binding site (NBS) resistance genes in pineapple. *Trop Plant Biol,* 9, 187-99.
[http://dx.doi.org/10.1139/gen-2012-0135]

Zhang, X, Yazaki, J, Sundaresan, A, Cokus, S, Chan, SW-L, Chen, H, Henderson, IR, Shinn, P, Pellegrini, M & Jacobsen, SE (2006) Genome-wide high-resolution mapping and functional analysis of DNA methylation in *Arabidopsis. Cell,* 126, 1189-201.
[http://dx.doi.org/10.1016/j.cell.2006.08.003]

Zhang, Y & Lewis, K (1997) Fabatins: New antimicrobial plant peptides. *FEMS Microbiol Lett,* 149, 59-64.
[http://dx.doi.org/10.1111/j.1574-6968.1997.tb10308.x]

Zhang, Y, Lubberstedt, T & Xu, M (2013) The genetic and molecular basis of plant resistance to pathogens. *J Genet Genomics,* 40, 23-35.
[http://dx.doi.org/10.1016/j.jgg.2012.11.003]

Zhang, Y-L, Zhang, C-L, Wang, G-L, Wang, Y-X, Qi, C-H, Zhao, Q, You, C-X, Li, Y-Y & Hao, Y-J (2019) The R2R3 MYB transcription factor MdMYB30 modulates plant resistance against pathogens by regulating cuticular wax biosynthesis. *BMC Plant Biol,* 19, 362.
[http://dx.doi.org/10.1186/s12870-019-1918-4]

Zheng, Z, Qamar, SA, Chen, Z & Mengiste, T (2006) *Arabidopsis* WRKY33 transcription factor is required for resistance to necrotrophic fungal pathogens. *Plant J,* 48, 592-605.
[http://dx.doi.org/10.1111/j.1365-313X.2006.02901.x]

Zhou, E, Jia, Y, Singh, P, Correll, JC & Lee, FN (2007) Instability of the *Magnaporthe oryzae* avirulence gene AVR-Pita alters virulence. *Fungal Genet Biol,* 44, 1024-34.
[http://dx.doi.org/10.1016/j.fgb.2007.02.003]

Zhou, M & Wang, W (2018) Recent Advances in Synthetic Chemical Inducers of Plant Immunity. *Front Plant Sci,* 9, 1613.
[http://dx.doi.org/10.3389/fpls.2018.01613]

Zhou, R, Yazdi, AS, Menu, P & Tschopp, J (2010) A role for mitochondria in NLRP3 inflammasome activation. *Nature,* 469, 221-5.
[http://dx.doi.org/10.1038/nature09663]

Zhou, Z, Ma, H, Lin, K, Zhao, Y, Chen, Y, Xiong, Z, Wang, L & Tian, B (2015) RNA-seq reveals complicated transcriptomic responses to drought stress in a nonmodel tropic plant, *Bombax ceiba* L. *Evol Bioinform Online,* 11 (Suppl. 1), 27-37.
[http://dx.doi.org/10.4137/EBO.S20620]

Zhu, Y, Chen, H, Fan, J & Wang, Y (2000) Genetic Diversity and Disease Control in Rice. *Nature,* 406, 718-22.
[http://dx.doi.org/10.1038/35021046]

Zhu-Salzman, K & Zeng, R (2015) Insect response to plant defensive protease inhibitors. *Annu Rev Entomol,* 60, 233-52.

[http://dx.doi.org/10.1146/annurev-ento-010814-020816]

Zimmerli, L, Stein, M, Lipka, V, Schulze-Lefert, P & Somerville, S (2004) Host and non-host pathogens elicit different jasmonate/ethylene responses in *Arabidopsis. Plant J,* 40, 633-46.
[http://dx.doi.org/10.1111/j.1365-313X.2004.02236.x]

Zimmermann, S, Nürnberger, T, Frachisse, J-M, Wirtz, W, Guern, J, Hedrich, R & Scheel, D (1997) Receptor-mediated activation of a plant Ca^{2+}-permeable ion channel involved in pathogen defense. *Proc Natl Acad Sci USA,* 94, 2751-5.
[http://dx.doi.org/10.1073/pnas.94.6.2751]

Zipfel, C (2014) Plant pattern-recognition receptors. *Trends Immunol,* 35, 345-51.
[http://dx.doi.org/10.1016/j.it.2014.05.004]

Zipfel, C, Kunze, G, Chinchilla, D, Caniard, A, Jones, JD, Boller, T & Felix, G (2006) Perception of the bacterial PAMP EF-Tu by the receptor EFR restricts Agrobacterium-mediated transformation. *Cell,* 125, 749-60.

Zipfel, C, Robatzek, S, Navarro, L, Oakeley, EJ, Jones, JDG, Felix, G & Boller, T (2004) Bacterial disease resistance in *Arabidopsis* through flagellin perception. *Nature,* 428, 764-7.
[http://dx.doi.org/10.1038/nature02485]

Zvereva, AS & Pooggin, MM (2012) Silencing and innate immunity in plant defense against viral and non-viral pathogens. *Viruses,* 4, 2578-97.
[http://dx.doi.org/10.3390/v4112578]

SUBJECT INDEX

www.ingramcontent.com/pod-product-compliance
Lightning Source LLC
Chambersburg PA
CBHW050824220326
41598CB00006B/307